LUCKY B-24

LUCKY B-24

Arthur Anthony

JANUS PUBLISHING COMPANY
London, England

First published in Great Britain 1993
by Janus Publishing Company
Duke House, 37 Duke Street
London W1M 5DF

**British Library Cataloguing-in-Publication Data.
A catalogue record for this book is available
from the British Library**

ISBN 1 85756 054 X

Cover design by Optigraph

Printed & bound in England by
Antony Rowe Ltd, Chippenham, Wiltshire

This story is dedicated to all the flight engineers who flew with Bomber Command in World War 2, particularly to a 20-year-old lying in Blickling Churchyard. His epitaph says it all: 'In his golden youth he took his place, a man among men'.

Acknowledgements

My sincere thanks to the Public Record Office for their co-operation in allowing me access to squadron operational records. My thanks also to Eric Munday of the Air Historical Branch of PRO; also to Len Bartram, Len Davies, Jim Bratten, Gerry Shutes, Hindryckx Didier, Bob Lawrence and Steve Correy. Finally, my thanks to Les Rogers for the introduction to PRO which has led me to writing *Lucky B–24*.

Contents

Foreword

Arthur Anthony's readable autobiography on his wartime career in the Royal Air Force centres on his operational tour as a flight engineer with a Liberator crew. His unit, No 223 Squadron, was one of those belonging to No 100 (Bomber Support) Group which pioneered the technique of providing radar jamming support over the target while the main force of bombers delivered its attack. In the recent war in the Gulf we saw the latest expression of these tactics, when EF-111 Raven or EA-6B Prowler aircraft accompanied raiding forces and jammed the radars of air defence missile batteries around the target. In each case the purpose of these operations was the same: to reduce aircraft losses and enable the attack crews to deliver their bombs with greater effect.

As the title suggests, Arthur Anthony drew strength from the luck that he felt attended him and his crew during his operational flights. For aircrews in time of war the perceived need to preserve good luck is much more important than might be imagined from watching the stiff-upper-lip actors that play aviators in films like 'The Dam Busters'. I talked to an RAF Wing Commander who led several Tornado attacks on targets in Iraq during the recent Gulf conflict. His talisman was a 'lucky wallet' that his grandfather had carried while an infantryman during the First World War, and his father had carried while with the RAF during the Second World War. Although a down-to-earth and brave individual, steeped in the technology of his aircraft, he confessed that he would have been very upset had he taken off on an operational mission and discovered that he had left the wallet behind.

Wars change, tactics change and aircraft change, but flying people hardly change at all.

Dr Alfred Price

Preface

What makes someone like me want to write about his wartime memories 46 years after V-E Day? Much has been written in the postwar period by various authors. Some of it by veterans who *were there*, and some by others who were not. The latter, at times, have let us veterans down by not taking the time and trouble to find out the real truth. Some of them have expressed ill-informed opinions which have swayed and influenced the reading public.

This is not a story of extreme bravery, just a simple account of a young man caught up in a war. I was badly educated, unaware of history, and poorly informed about everything in general. My two best subjects at school were spelling and composition. Together with my memories these assets have taken root and bloomed into *Lucky B–24*.

Arthur Anthony
Colchester, Essex, 1993

1 RAF Volunteer

I had just turned sixteen when war started on September 3rd, 1939. Up until that fateful date I had seen only one German. Most of us had no idea of the preparations being made by both sides for another world war.

My father had been a sailor in the Merchant Navy before and after World War I. During that war he served on troop, munition and hospital ships in the Dardanelles campaign. One night he fell from the crow's-nest whilst keeping look-out in convoy. Although he was badly bruised not a bone was broken. Such a crash to the deck could have killed him – he was lucky to survive. When it came to luck I was to follow in his footsteps. He automatically assumed I, too, would become a sailor.

Back in those days one was called up at nineteen years of age, or could volunteer at eighteen. I had no desire to become a soldier or a sailor, particularly after listening to 'old sweats' who had served in the trenches. So two weeks after my eighteenth birthday I volunteered for the RAF. I was accepted for training as a flight mechanic, engines (FME) or airframes (FMA). I had fancied myself as aircrew, perhaps as an air gunner, but at this stage of the war one had to qualify as a wireless operator (Wop) as well as a gunner. Morse was not for me.

2 605 Squadron, Ford

October 1st, 1941, found me at Cardington, Bedfordshire. We were kitted out and sent to Great Yarmouth to do our square-bashing. Whilst there, Jerry was a regular visitor. We lost all our underclothes when the laundry was bombed. We all had clothes that didn't fit.

It was the usual things at Great Yarmouth: jabs, physical training, drill and 'bags of bull'. I fired my rifle in an ex-funfair where, in 1939, I had enjoyed so much pleasure. Halton in Bucks was a technical training school, and was so full of trainees it could not absorb us, so it was a further three weeks of 'bull' at Great Yarmouth.

Luck played a major part in one's service life. My first piece of it came when I managed to be selected as flight mechanic, engines. After sixteen weeks of technical training at Halton we 'passed out', and our pay was increased to about six shillings a day.

I was posted to 605 Squadron at Ford, Sussex, in the early summer of 1942. At that time 605 Squadron consisted of half a hut which was the orderly room. The squadron had been wiped out by the Japanese in the Far East and was being reformed at Ford with Havoc Is and Boston IIIs. These aircraft were inherited from 23 Squadron, a night-intruder outfit which was soon to leave Ford re-equipped with night-fighter Mosquitoes.

About this time Daily Routine Orders were asking for volunteers from the ranks of FMEs and FMAs to train as flight engineers. This rather appealed to me, so I put my name down, only to receive a mass of gibes and jeers from a lot of the 'old sweats', such as, 'Are you tired of living?' and 'You must be bloody mad'. Quite frankly, I had no idea of the casualty rate in Bomber Command, as the powers that be were not very forthcoming on this point. It was the chaps who had got some service in who had seen the crashes, and they'd seen kites off that never returned. Hence their rude remarks about volunteering for aircrew.

Every aircrew trainee in the RAF was a volunteer; this was not so in many other air forces. I was soon aware of what they were getting at over the next few weeks when 605 Squadron started their training programmes, readying themselves for night-intruder operations.

I was anxious to get airborne and get some flying experience. I managed to do so one afternoon. It was great until the pilot, Pilot Officer C. F. Ponsford, threw the Havoc all over the sky; he followed this up with dives and turns and then low level stuff. It was not very long before what I'd had for dinner was spread all around the navigator's compartment. To be quite honest I was more than glad when we landed and arrived back at the dispersal point where I was still feeling pretty rotten. As I got out I pulled the emergency lever on the escape hatch by mistake. They were quite a job to reassemble. Feeling a right fool, I picked up my parachute by the *D* ring; next thing I knew, it billowed out and was blown everywhere. I had to face an awful lot of taunts from my fellow airmen:

'You silly sod, you will be useless as aircrew'.

To cap it all it cost me half-a-crown for someone to clean up the remains of my dinner. Two shillings and sixpence was an unwritten standard charge in the RAF to pay a bod to clean up a kite after air sickness. I got more than my share of ribbing over this episode and felt the only way to live it down would be to get airborne again.

Within a few days I managed to do this as there was a busy training programme. This time I flew in the gunner's position which was much more to my liking than that first flight in the nose. After a lot of dummy attacks and evasion manoeuvres with another Havoc we landed. I had every intention of going up again and waited whilst both planes were refuelled. At this juncture three Canadian soldiers appeared on the scene. There were lots of Canadians in Sussex at this time of the war, waiting to go on that ill-fated attack on Dieppe (to which I will soon refer in more detail). All three of them asked if they could have a trip up. I advised them to ask the pilot, pointing out that there would be plenty of other chances for me to fly another day. So off they went, three Canadians and three RAF aircrew, the former all smiles.

It was not long before my 'chiefy' (flight sergeant) spotted me and wanted to know what I was up to. He quickly gave me orders that the two aircraft, when they returned, would be my responsibility. I was to refuel them both. It was getting a bit late in the afternoon and I was worried that I might miss my tea, but chiefy promised me a chit to get a late tea. A chit in the RAF was a very valuable commodity and could get one lots of concessions and perks. I suppose less than half-an-hour had gone by when 'chiefy' came over on his bike from his little flight dispersal hut.

'Go and get your tea, Anthony,' he said.

'But what about refuelling these kites?'

'You won't be doing that, I've just had a phone call to say that the aircraft have had a mid-air collision over Bracknell and all six are dead. You're a lucky little sod.'

I certainly was; those Canadians had saved my life but lost their own. I was forever lucky for the rest of my service career. One of the dead aircrew had been with 605 Squadron for just two days.

Within a very few weeks Lady Luck smiled on me again. This time, whilst servicing a kite one morning which had been on a night-intruder patrol, I and the rigger had pulled the port engine through and had just crossed past the nose to do the starboard when the four Browning machine guns were fired accidentally by the armourers testing the electric firing solenoids. It was a close thing. Both of us had missed death by a few seconds. There were 22 lifeless .303 cases laying on the ground, and it was just luck again that we were not lying there with them. Later on in the day, whilst testing the engines, I switched on the reflector sight in the cockpit which suggested that the 22 bullets had travelled towards Arundel Castle – probably the only time during the war when it had come under fire. Of course the 'chiefy' of the armoury section played up hell about it and was threatening 'charges'. We never did hear where those bullets finished up. No charges came from this episode but within days the armoury 'chiefy' blew his top again when a 20 mm cannon shell went through the top of his dispersal hut. Some armoury bods got seven days jankers for that little episode.

Wing Commander Peter Townsend was the commanding officer of 605 Squadron. A Battle of Britain veteran, he was later to become Equerry to King George VI. It was at the Palace that he met Princess Margaret whom he wanted to marry, but the relationship did not bear fruit. The squadron CO carried out the first 'ops' on 14th July 1942, when he left Ford at midnight with two 250 lb and twelve 40 lb bombs, before he released them on railway sidings west of Mandeville Power Station – a round trip of 2 hours and 11 minutes. By the end of July we had six Havocs and 14 Bostons.

Ford was a very busy airfield with lots of activity. One day we had a photo reconnaissance unit (PRU) blue-painted Spitfire crash land and catch fire, but the pilot got out OK. The squadron lost its first aircraft on ops on 2nd August, 1942, when Squadron Leader J. D. Humphreys failed to return from Beauvais, France, carrying two 250 lb instantaneous fused bombs. Humphreys had already attained a fine war record. He had flown 127 operational night sorties. Fifteen days after his plane went missing his body, together with that of his navigator, was washed ashore at Brighton. Their tactics were to fly into France at very low level to avoid tracking by German radar. Not a very easy task to carry out in darkness, and mistakes did happen. We heard on the grapevine that part of his seat was still attached to his body which suggested to us that he hit the water and exploded his two bombs.

On 7th August Peter Townsend was posted to Fairwood Common and Wing Commander G. L. Denholm took command of 605 Squadron. He was a very experienced pilot and had flown with 603 Squadron as a squadron leader during the Battle of Britain. On 14th August we were sent to Hunsdon near Harlow. We left behind our two flight commanders, K. Sutton and M. G. Olley, who, together with 418 Canadian Squadron, also with Havocs, took part in the Dieppe raid on 19th August when they attacked gun positions. Two or three enemy aircraft attacked Ford whilst the Dieppe raid was in progress. They dropped four bombs which badly damaged two Bostons. We all thought Jerry was a bit cheeky, even more so when a few days later we picked up German propaganda leaflets showing pictures of smashed boats and tanks and long columns of Canadian troops taken prisoner, with their hands on their heads.

We returned to Ford the next day to find a large number of kites lying about, all badly damaged. There was another fatal casualty on 28th August when P/O Pierpoint and crew, on a night-flying exercise, were caught in searchlights and crashed between Bognor and Ford. They were all killed instantly. We had received an Owlet light training aircraft on 2nd July; it was very useful getting pilots used to the experience of a tricycle undercarriage. The aircraft only lasted 60 days. It crashed close to Arundel at 5.15 p.m. on 31st August, killing both the pilot, P/O G. M. Jordan, and his navigator, T. A. Dunphy. We also acquired a new addition to the squadron, Lieutenant Colonel Kratz, a USAF attachment for liaison purposes.

It was around this time that we had our first serious accident to a member of groundcrew. It happened while I was on night shift.

We had several crews on night training exercises. Most of us were in the flight dispersal hut when someone rushed in crying out: 'For God's sake come quick, a chap has been hit by the props'. It was my very first experience of seeing so much blood; he was screaming and crying out. It took about six of us to pin him down; he was a strong, well-built amateur boxer. An officer gave him a shot of morphine from out of an aircraft's first aid kit. The poor WAAF driving the blood wagon was in a shocked state, so I jumped aboard and guided her back to Station Sick Quarters. It was quite some time later that we learnt that the badly injured lad had survived and had a silver plate inserted in his head.

Ford was a busy station to serve on and we had lots to keep us busy, seeing to visiting aircraft as well as our own. One morning, going on duty, our first job was to help clear the runway of a 'wheels-up' landing. A 'Wimpey' (Wellington), had been hit by night-fighter cannon shells. The rear turret gunner's blood had blotted out any view into the turret. I congratulated the Australian pilot on his fine belly landing. I shall always remember his answer:

'You mend them', he said, 'I'll bend them'.

There was a big heap of undropped propaganda leaflets in the fuselage, even at this stage of the war.

On 2nd October 1942, 605 Squadron were again pushed out of Ford, this time to Debden. This was so Bomber Command could use Ford for large-scale daylight offensives on 'circus' operations* with Bostons, etc. We used Castle Camps, near Haverhill, for a couple of nights for night landing training, but after a week we were back at Ford. On 21st October one of our Boston IIIs was on a night training exercise when it had a mid-air crash with a Hurricane from 534 Squadron based at Tangmere near Chichester. Both crews were killed.

It is interesting to note that Bomber Command did not have a monopoly on leaflet dropping. On 16th November 1942 we sent out our Bostons loaded up with propaganda leaflets for the Belgians to read. Most of it referred to Monty's victory at El Alamein. It was called 'nickelling' in official circles.

Around this time Flight Lieutenant Williams was posted on a pilot's course; he had flown with 264 Squadron – Defiants – in the

* Circus operations: daylight bombing with medium bombers, heavily escorted by fighters.

Battle of Britain and had won a DFC. Squadron records showed that he was a professional parachutist with 400 jumps to his credit.

During October we were kept very busy, not only tending to our own aircraft, but also seeing to Lancaster and Halifax bombers which lobbed into Ford after long trips to and from Germany and Italy. Sadly a Lanc crashed off the runway coming in to land after a daylight op on Turin. The Press next day hailed it as a 'White gloves raid', because there were no losses. The poor devils who died in the Lanc never got a mention. Within a few days I was detailed for the funeral party which was a very sad occasion. Four years later I was to meet the station warrant officer (SWO) of Ford at Abingdon. He gave me a photograph of the funeral, mentioning that he took photographs of every RAF funeral in Littlehampton cemetery.

Littlehampton was affectionately known as *Little Chopper*; this wording came from the many Fleet Air Arm airmen who had a small servicing unit, which, amongst other things, dealt with Walruses and Sharks.

Towards the end of 1942 *A* Flight received a new commander, Squadron Leader C. D. Tomalin. He had been an Olympic Games diver.

Sergeant Pilot Warrender did not see in the New Year. He and his crew were killed when his aircraft, UP-Z, a Boston III, crashed at Kingston-by-Sea on New Year's eve. They had been on an intruder patrol to Chartres. An enquiry board later stated that the pilot had ignored his altimeter and had hit the ground on approach. At this time we had two problems:

- A shortage of ground crews and poor working conditions for those we had.
- Too few buildings caused by a shortage of roof trusses.

We had no fewer than nine Lancs to see to on 22nd January 1943; they had been to Düsseldorf. We also had 264 Squadron's Mosquito Mk 2s on our hands. They had flown in from Colerne, Wiltshire, to Ford but did not operate. One morning I was on a Lanc's wing, checking oil levels, when a Beaufighter lined up for take-off and began to speed down the runway. Watching its progress I knew instinctively that it had been pulled off too soon and watched helplessly as it stalled and dropped back on to the runway. It slewed all over the place and for one moment I thought it would plough into the parked bombers. I never did remember getting off that Lanc wing but I know it was in a hell of a hurry. I found myself running towards the now stationary wreck along with many other bods. The pilot was already out of the aircraft followed by the radio observer.

Someone asked him amidst the tense atmosphere, 'Are you okay?'

'Yes', came the reply, 'but I think I have sprained my wrist'.

This was greeted by howls of laughter by those who had run to the aircraft's aid. There were petrol, oil and 20 mm cannon shells strewn everywhere. Both oleo legs had been sheared off; how it failed to catch fire remains a miracle.

That word *luck* played such a great part in the war. I later learned the cause of the premature take-off and stall. It would seem that an electrician servicing the runway lighting had walked in the path of the Beau as it came down the runway. The pilot had pulled back on the stick and literally 'hopped' over him, hence the crash.

Walking across runways unofficially was a deadly sin in the RAF but I had seen the CO, Wing Commander Denholm, do it on more than one occasion. This was to lead to a bit of fun I had at the CO's expense. When the night duty crew had no aircraft to see off, usually because of bad weather, we had to patrol the dispersal area with a rifle and 50 rounds of ammunition. There had been complaints that we were not carrying out this order, for even at this stage of the war German parachutists were still expected to make commando raids on airfields.*

One evening I saw the wing commander leave the Control and Intelligence block with his navigator. I was on sentry duty and in the gathering dusk saw him, once again, cross the runway, short-cutting to our dispersal. I hid up close to an aircraft and let them get close to me. Then I jumped out and thrust my rifle at Denholm's chest.

'Halt! Who goes there?'.

'Friend', he cried.

* I am still of the opinion that, had this been attempted, terrible havoc could have been wreaked amongst parked and bombed-up aircraft. There are many of us who today still consider that a more intensified intruder campaign by the Luftwaffe would have given them much better results. Of course we now know that this was a German intention, but Hitler, as usual, put in his 'four eggs' and virtually put a stop to it. Hitler wanted the German public to see destroyed Allied aircraft fall on German soil.

Poor devil, I really scared him. Later the night duty corporal told me that the CO had called in at the dispersal and said, 'You have a strong guard on here tonight, Corporal'. I told him what I had done. We never got any more complaints about sentry patrols from Denholm.

Bob Gowing, a rigger and great friend of mine, saved the CO's life only a few days after the sentry incident. He placed the CO's chute and dinghy pack in the cockpit prior to an intruder operation. A few seconds later the CO^2 bottle malfunctioned and inflated the dinghy. It was fortunate this happened when it did and not whilst he was airborne.

3 Preparing to Become a Flight Engineer

Eventually I stood in front of Wing Commander Denholm for an interview with regard to my becoming a flight engineer. I flannelled a lot of his questioning:

'I see you like horse riding?'(I didn't tell him I had only done it on one occasion.) 'Could you sit in a bomber for nine hours?'

For a moment I was speechless. Then I answered, 'Well, yes, of course I could, if others can do it so can I'.

I must have impressed him because he said he would recommend me for a medical and selection board in London. This came up and I was accepted for training.

On 25th January 1943 Flight Lieutenant Olley took off at 12.30 hours with his crew and scattered the ashes of the late Flight Lieutenant Richard Hillary, author of *The Last Enemy.* A memorial service was held in London at the same time as the ashes were released from the aircraft. Just 44 days later Olley did not return from an intruder patrol to Chartres and Tours, flying the very first ops on 605 Squadron's new Mossies. Olley had given over 2000 hours of dual instruction. The squadron had lost a great man who was liked by everyone.

We were paid once a fortnight. The old hands referred to it as 'When the golden eagle shits.' Squadron Leader Sutton, one of our flight commanders, always seemed to get airborne on pay days and would buzz the pay parade. I always thought this New Zealander was showing off. We did get a good laugh on one of these 'Golden Eagle' occasions when he took off and came back in and landed. Why such a quick return? We saw why when his navigator stepped out of the nose; he was in his best blue, ready for going on leave; he was smothered in blood and feathers from a seagull which had penetrated the nose canopy on take-off. Later, flying a Mossie, the same pair came in at low level over the sea somewhere on the South coast. Our own flak opened up on them and killed the navigator. Sutton managed to get past his dead body which was miraculous. He landed by parachute but he lost an arm in the process. He must have been a tough nut. The last I heard of him he was back flying Martinets at a gunnery school in Wales.

At the end of January 1943 I was posted to Cosford to do a Fitter 2 E's course. At this juncture of the war flight engineers had to be qualified fitters, so off I went for another 18 weeks in technical training school. After this I assumed it would be my FE's course. Not so; I was posted back to my old 605 Squadron who were now operating from Castle Camps near Haverhill with their night intruder Mosquitoes.

I looked forward to the night shifts. Why? Well, early on there was not a lot to do, so I would slip off and set my rabbit snares amongst the barbed wire perimeter defences. After seeing the kites off and refuelling on return we could get our heads down. Overnight I would often catch a rabbit and nip home next day with it, either by bike, train or hitchhike. My parents were always pleased to see me, and even more so with that day's dinner. I got a lot of ribbing from my billet mates. 'Lucky b . . . posted on his back door step, and got his bloody dinner to take home as well.' But the boys were always pleased to see my return from home with a tin of dried milk powder that my mother had managed to come by. We used it to make tea in the billet. We could get our hands on tea and sugar, but milk was the problem.

I had another bit of luck whilst at Castle Camps. Around teatime one day we got an order, 'Bad weather, no flying tonight, cover up engines'. I soon had both engine covers on and tore back to the billet to pick up my irons en route to the cookhouse. Just then, over came an FW 190; he dropped a bomb right into our dispersed Mossies and set on fire the one I had just covered up. A piece of the bomb shrapnel had hit the wing long range auxiliary tank. Half a wing was gone before the flames were subdued.

We had quite a lot of trouble with intruders at Castle Camps. It was literally intruders intruding on Intruders. One night the WAAF duty driver, whilst in her bed, was blown by blast through a Nissen hut, but she was unhurt. Another night I saw a Stirling catch a packet of tracer. We heard that it was crash-landed on fire, short of the runway, at Stradishall, Suffolk, which is now known as

Highpoint Prison. Strangely enough, my old base at Ford is now also a prison.

There was a strong rumour that we were getting intense intruder activity because the Luftwaffe pilot had been to Cambridge university and knew the Cambridgeshire area very well.

Training at St Athan

I was posted to St Athan, Wales, at the end of September 1943 to do my Flight Engineer's course. Just a few days after I left Castle Camps, 605 Squadron, with their Mossies, moved to Bradwell Bay, Essex, which was to become the home of a nuclear power station. One usually had to wait a week or so to get started on a technical training course, and stayed in what was known as Pool. You could get allocated all kinds of rotten jobs serving in Pool. I knew a thing or two by now and volunteered straight away for the first job that came up. I did just this at Cosford near Wolverhampton and got a job guarding a would-be escapee in the camp IDH (infectious diseases hospital). It was there that I learnt all about the perils of VD and also found out what a scourge it was. Keeping an eye on a bod was much easier than heaving coal in the camp coalyard.

St Athan was a huge prewar RAF camp which, at this stage of the war, was dedicated and geared up to producing flight engineers. Its total of over 20,000, including early postwar trainees, is a fine record. Direct entries from civvy street took 26 weeks to train. Fitter 2 E's and A's took just seven weeks. I served a week in Pool; I got a nice little job down at the airfield carrying out some simple work on their one and only B.24 Liberator. I had not seen one on the ground up to this date, and was impressed with its size, comfortable seats and robust looks; so began a love affair with it.

A sort of pecking order governed the way in which trainees were allocated to types of aircraft at St Athan. There were a few vacancies to train on Yorks, Sunderlands, Catalinas, and Fortresses. Corporal and sergeant fitters had the monopoly of this group. Then we got to Liberator allocation.

'Fall out all those who want to train on Liberators,' the NCO called. He added, 'It will mean an overseas posting to a bomber squadron.'

That did not deter me or the other 200 odd would-be trainees.

'We only require 40 men, so there will be a process of elimination.'

Direct entry were the first to be shoved off. Next it was all those under 6 feet tall. I was still there, along with a dwindling but still too large group of Liberator enthusiasts.

'All those whose serial number on their 1250s (identity cards) ends in 0.'

There it was on my 1250; I showed it and my name was on the UT (under training) Liberator list. Lucky me, I was getting my share of it. They went on then to allocate people to Lancasters, Halifaxes and Stirlings. Nobody wanted Stirlings, but got them just the same.

Several blokes threw their hands in the next day. There was a lot to learn; I found the going pretty hard, especially the maths on fuel consumption. I often think how nice it would have been to have had one of today's pocket computers. The course included an hour's dinghy drill, and another hour dedicated to ground parachute training. It is surprising the number of times the word 'coincidence' cropped up in one's service life. I met three lads at St Athan who had all been at school with me. All three were direct-entry trainee flight engineers (FEs). Their fate is recorded in later pages. At the end of the seven weeks' training I had failed the course, but only on one subject, the fuel system. I was given a further week's training and was home and dry.

We were sent on leave, to return to St Athan just before Christmas. Whilst at home I received a telegram; it read: 'Your leave is extended until the New Year'. I couldn't believe it. I have said in previous pages that I was prone to Lady Luck, and it had happened again. I had not had a Christmas at home for three years. It was during this leave that I met my future wife. After FE's school I was sent to Whitley Bay, Newcastle, for an NCO's course. It consisted of an assault course, grenades and sten guns and 'bags of bull'. After three weeks of living in private houses taken over by the RAF we were posted back to St Athan. Before leaving, and to celebrate, several bods drank far too much and ran amok with their bayonets until they'd done in all the internal doors of their billet. They faced a court martial but the CO was a decent type – he let them off with a caution and heavy barrack room damages.

The High Seas – Destination Unknown

Soon we found ourselves in Blackpool being kitted out for overseas. I received my kit from a WAAF whom I had known at Ford. She was very unhappy – her boyfriend had lost his ears in a blazing flying accident. At this stage none of us knew our overseas destination. Early one misty and cold February morning found us leaving a train at Greenock, Scotland; we were marched on to what I thought was a rather small boat for the high seas. I have mentioned before that I had no sailing instincts. Enquiring of a hugely built US coastguard, he pointed and said, 'This is just the tender, your ship is out there'. Soon, out of the mist, loomed a huge ship; it was the *Queen Mary.* After being allocated quarters we were all given different jobs to do. Later that evening we sailed for New York. Next day we were going like hell on a constant zig-zag course with no escort of any kind. There were all sorts on board but the *Mary* was not crowded – there was plenty of room for everyone. There were WAAFs en route to Washington, a few ATS girls and a handful of soldiers. There were also quite a number of airmen. Royal Navy men were on their way to pick up and crew American-built ships and sail them back to England. Some of the sailors were allocated to the rear gun on the *Mary*; they didn't even know how to load it. Later that night one of the sailors was found asleep at the gun and was arrested. He spent the rest of the voyage in one of the ship's cells. The *Queen Mary* had done quite a few voyages to South Africa as a trooper, and for the first couple of days of her voyage she ditched her huge stock of cheap and nasty South African cigarettes on us. They were aptly named 'C to C' brand, which stood for Cape to Cairo. After smoking them it seemed more likely to mean 'Cough to Coffin'. What were left of them were thrown overboard, aimed at the military band and their instruments as they played a welcome home for the American servicemen on board. These were mostly USAAF aircrew who had completed their tour of 25 operational sorties. Many of them wore Purple Hearts.* I was to learn that one merited one of these if one got frostbite.

* The US Armed forces medal of the Order of the Purple Heart, awarded to servicemen wounded in action.

Trains and Boats

We were allowed a few hours for sightseeing in New York and managed to get to Times Square and the Stage Door Canteen. This latter establishment was sponsored by film stars and provided canteen and entertainment facilities. I was not that impressed; the large smoke rings blown across Times Square by the Camel cigarettes advertisement drew more of my attention.

We had to be back at Grand Central rail station by 8 o'clock in the evening. On the way there the air raid sirens wailed and there was a 'blackout'. I am sorry to say there were scenes of panic. It was only a practice job. We had quite a struggle getting to Grand Central on time, the policemen trying to make us take cover on the way. We were more intent on catching our train.

Our next move was a train journey to Lachine, SW of Montreal, Canada. There we found out that the French Canadians were not all that fond of us. One of them tried his hardest to pick a fight with two of us one evening, but we managed to keep our cool despite all the insults. We just had to – he was carrying a cut-throat razor. Montreal was a busy place – lots of places to eat and drink – but it was the food that appealed to me. After nearly four years of English wartime food the bread was so white and tasted like cake. The ice cream, chicken and steaks were out of this world. Just a week of this new-found heaven and we were on the move once more. We had a long train journey ahead of us to Miami, Florida. It took about three days on the train to get there. We left Miami on a small supply ship, the *Jean Brilliant*. She was crewed by French Canadians and was about 350 tons. Only minutes after moving into the open sea at dusk the ship hit a severe storm. Conditions were made worse as the vessel was out of trim. Her deck cargo of large wooden cases containing aircraft engines were all loaded in the bow. We were all seasick including some of the crew. They had never experienced such a rough crossing. Nassau in the Bahamas was our next stop. Next morning on deck it was beautiful, a lovely sparkling sea and blue sky, with flying fish jumping in the path of the ship.

Over the years press articles have dealt at length with old 'favourites' of theirs, the Duke and Duchess of Windsor. Referring to his appointment as Governor-General of Nassau they have dismissed the place and his job as insignificant. His job may have been so, but the columnists were ill-informed in referring to Nassau itself in those terms. It was to play a major role in the winning of the Battle of the Atlantic. Early in the war Churchill gave President Roosevelt concessions to build two airfields on Nassau in exchange for certain naval rights. 111 OTU Nassau provided nearly all the trained crews of Coastal Command Liberators which, with their extremely long range, were able to close the vital gap in the Atlantic defences, and thus greatly reduce the U-boat menace.

111 OTU was opened on 30th December 1942. Its CO was Group Captain Waite, who, later in his career, was to mastermind the Berlin Airlift. Just one year after it opened, Nassau had an establishment of 2000 staff, plus 750 pupils, who included elements of the Czech, Polish and Dutch Air Forces. It consisted of two fields: Oakes, where the B.25 Mitchell was based; and Windsor field where the Liberators were flown, along with elements of Transport Command engaged in moving new aircraft from the States (such as Mitchells, Marauders and Dakotas) via South America to the Middle East or UK.

My first four weeks were spent at Oakes learning air gunnery on the ground, plus Morse and aircraft recognition. It was at this time that we were allocated crews who were at this point flying the Mitchells. Once again, a 'pecking order' reared its ugly head. A list of trainee captains was posted up; one's turn in choosing depended upon the amount of service one had 'got in'. I finished up with a Dutchman for skipper and an ex-operational air-gunner as co-pilot. I was quite happy about my selection of captain, but a few days later it all went terribly wrong. One of my fellow flight engineers asked me if I would swap crews with him. I asked why he wanted the change. The reason, he said, was because there was a chap in my crew who lived in the same town as himself and it would be nice to travel home together, on leaves, etc. I was a stubborn so-and-so in those days and refused to change crews with him. He made several more attempts to talk me into swapping, including getting a few of his friends to plead with me. In the end I lost my temper over the subject and at last agreed to change. I did not know it, but fate was dishing out more favours to me. The crew I parted with didn't make it to the end of the war. One of the reasons I didn't fancy changing crews was because his pilot was a bit posh.

At Last the Real Thing

Moving up to Windsor Field we began our training on the B24 Liberator. Nassau was an ideal choice of place to train aircrews. Crews could fly around the clock, and the weather seldom interfered with the flying programme. I did myself a bit of good changing crews. P/O Briscoe, my new pilot, was 32, the maximum age for training pilots. Roxy Hart, the co-pilot, was just 20 which was average for the rest of the crew. Briscoe was a fine looking, well educated man. He had been to school at Harrow, and to Oxford University where he had qualified for his wings with the University Air Squadron. His father, Sir Charlton Briscoe, was an eminent EN&T specialist in Harley Street, London, and the family lived at Lakenheath Hall, Suffolk. Roxy was not in the same class as Briscoe when it came to flying, yet he was a very likeable person and hailed from Little Clacton. That made three of us in the crew who were East Anglians. Flight Lieutenant Merrifield was our initial instructor at Windsor Field. He was to have a long and successful postwar career with British European Airways and became Chairman of BALPA. After just a few exercises at landings and take-offs he handed the stick over to Briscoe. It was now that I was able to get in my air gunnery practice and fired the .5s from all the defensive positions. Up until now, not a Liberator had been lost from Nassau. There had been crashes with the Mitchells resulting in fatalities. Briscoe let me sit in the co-pilot's seat and encouraged me to take the stick and have a go at flying our Liberator. That gesture made me feel about 8 feet tall.

In no time at all the final 'Kingsley' exercise day arrived. This entailed flying with a full petrol load, carrying eight Torpex depth charges to a rendezvous with a British frigate operating out of Bermuda. It involved a four-hour flight from Nassau. The navigator insisted we should be able to see the frigate at any moment, but there was not a sign of it. The radar operator could not find it on his screen. Eventually the staff pilot pushed him to one side and quickly homed in on the ship. When we made our first low and slow pass over this Royal Navy vessel I could see lots of matelots sunning themselves on deck. Shortly after our arrival the frigate put down a smoke float which we 'attacked' at zero feet with one of our depth charges. As we climbed and broke away I had instructions to open fire on the smoke float from the port waist gun. I must have got carried away with it all because I could hear the skipper calling on the intercom to 'cease fire'. I then realised that my shooting was ricocheting and my tracer fire was crossing low over the ship. On our next pass over the ship there wasn't a soul left sunbathing – I had literally 'cleared the decks'. The navigator had also boobed a little bit. He had dropped two depth charges instead of the scheduled one. However, I did come out fairly well on my fuel consumption figures; when tanks were dipped after nine and a half hours' flying, their figures were very close to my own.

Tragedy had struck for the first time on this particular Kingsley exercise. A Liberator was missing with all its crew, together with difficult to replace instructors on board. Quickly a large scale search was made of the whole area. No distress signals had been picked up so it could have been anywhere in that vast expanse of sea. Despite many hours of searching by US Naval Air Forces, Royal Navy and units from 111 OTU, no one was found. In September 1989 I went to San Diego to celebrate with other veterans the 50th anniversary of the building of the first of 18,000 Liberators. Whilst there, I got into a long conversation with an ex-RCAF and Air Canada pilot who had 20,000 flying hours to his credit and learnt that he had taken part in the sea search for the missing Liberator from Nassau. Flying a Mitchell, his crew thought they had found a dinghy and could see something white moving about. They dropped down lower to investigate as it was a very sparkly sea. Their high hopes were dashed when, on closer inspection, their so-called 'dinghy' was just an overturned dead whale with a lot of seabirds on it busily eating the carcass. After I left Nassau a further two Liberators were lost when they had a mid-air collision. As the Kingsley exercise brought an end to our training at Nassau, we all celebrated on the Saturday night. I think I had a little too much to drink and passed out in some hotel. The RAF Police came in looking for a drunken airman. I was told afterwards by my crew that they all sat on a settee with me full-stretch at the back of them, and then they smuggled me out to a waiting taxi that took me back to camp. I never did pay the skipper the eleven shillings he forked out to the taxi driver, but I did offer. As the stay at 111 OTU drew to a close our skippers asked their crews their preference. We had a choice of posting – the Far East or back to the UK. I plumped for the UK. I was longing to see my girlfriend again. The rest of the crew also voted for home. Lots of the Australian air crews wanted Far East postings, so there were no problems on this point.

UK Bound

We sailed away from Nassau after a delay; we were in the same little boat that had brought us there. She had a problem with her steering gear and Morsed the fact to her little American naval escort which was patiently waiting outside the harbour for us. It was not, incidentally, yours truly who read the Morse message. At last it was goodbye to Nassau. We had not seen the Duke of Windsor nor his Duchess but we had sampled the nice Services Club that she had promoted and helped to run. After one night in a hotel right on Miami Beach we embarked on another long railway journey back through the States to Moncton, Canada. We had to wait a further three weeks before we left for the UK.

I went to a camp dance one evening and again that word *coincidence* cropped up. Within a space of five minutes I met two people I had known back home. The first was a sergeant PTI who had been a policeman in my home town of Colchester; I had got to know him whilst boxing for the local Home Guard team. Colchester Borough Police Force was renowned, prewar, for its boxing team. Most of the boxers were recruited from the local army establishments in the town. If one was a good boxer in the army it was a ticket into the local police force. Jock confided that he had recently returned from Madison Square Garden where he had taken part in a boxing tournament. After the war he became a pro but did not last too long when Bruce Woodcock whopped him a good one. The other Colchester chap lived just across the road from me; he was a sailor en route for the UK after his naval pilot's training had been discontinued. The navy had a large surplus of aircrew and were sending him back to civvy street as he had been in a reserved occupation.

Soon we left Halifax, Canada, on board the *Empress of Scotland* heading for England. We were packed in like sardines. I was right down in the bowels of the ship, sleeping (or trying to) in a hammock. It took a week before one got a decent night's sleep, but before that had elapsed we arrived back in Greenock. For the first few hours of our voyage we had received a RCAF escort but after that it was the same style as that of the *Queen Mary* – full out and zig-zag all the way.

It was lovely to hear the army band play us out of the railway station on our way to Harrogate. I was looking fit and tanned, despite those awful nights in a hammock; I would never have made a good sailor. I had certainly been a long way in the last 14 weeks. More like a holiday, some would say. We had only been at Harrogate a few days, waiting to go on disembarkation leave, when early one morning an NCO came on his usual tour of duty getting us out of bed:

'Come on chaps, out you get, the invasion has started'.

At first I could not believe him. We seemed to have waited so long for it to happen. After two weeks' leave it was back to Harrogate. This Yorkshire health resort was chock-a-block with NCOs and officers who had returned to the UK from various training schools in Canada, USA, South Africa and Rhodesia. Squadron Leader Leslie Ames was the commanding officer. Some of my readers will know that he was a well-known cricketer before the war, not much liked during the same, and not a very popular man. I was put in front of him on a charge for being AWOL. He could not make the charge stick as I was able to tell him who had called the roll on that particular occasion. We were all kept well occupied with route marches, shooting skeets and a lot of 'bull'. The days dragged on but a posting to an operational Coastal Command squadron did not transpire. We wondered if we would ever fly on ops.

4 223 Squadron, Oulton

Then out of the blue, came a posting with 15 other Liberator crews and we were on the train to Oulton, Norfolk. There was lots of time en route to discuss and speculate about the new posting. The pundits amongst us were adamant about one thing: Oulton was not a Coastal Command airfield. It had been built on part of the Blickling Hall Estate.*

Most of the officer aircrew of 223 Squadron (to which we now found ourselves attached) were billeted at Blickling. The commanding officer had the honour of sleeping in the bedroom once occupied by Ann Boleyn. Like most significant airfields, Oulton had a satellite. Oulton's satellite was Great Massingham airfield, $7^1/_2$ miles WSW of Fakenham. In the tearoom at Blickling there is a mounted plaque which bears a list of captains who served with 223 and 214 Squadrons. This token was presented by the widow of Group Captain T. C. Dickens who was the Station Commander at RAF Oulton during my time there.

It seemed ages on the train with lots of stops including one at Lakenheath, where the skipper shook hands with the stationmaster. It seemed that everyone knew the Briscoes from Lakenheath Hall.

* Blickling was $1^1/_2$ miles from Aylsham and 15 miles north of Norwich. It is now owned by the National Trust. Built of red brick early in the 17th century, it sits in beautiful parkland. It had Dutch gables and contained a stateroom in which hung many fine paintings. In the gallery there is an unusually long Jacobean plaster ceiling. The gardens are laid out to a late eighteenth-century design. The park contains a crescent-shaped lake and a mausoleum. There is also an orangery, and a temple by Ivory of Norwich. Sir Henry Hobart was the original builder of Blickling Hall, a Lord Chief Justice and one of the first baronets to be created by James I in 1611. Later the Hall came into the hands of the Marquises of Lothian, and the 11th Marquis bequeathed the estate of 4,500 acres, plus the Hall and its contents, to the National Trust when he died in 1940 whilst Ambassador to the USA in Washington.

Author: Arthur Anthony

Sir John Briscoe DFC. The Skipper

Sir John's crew

Vic Green & Arthur Anthony

Duchess of Windsor Services Club

Aerial view of Windsor Field, Nassau

Rendezvous with frigate out of Bermuda

Burial of Lancaster crew, Littlehampton

Flight-Engineers of 223 Squadron

F/Lt Morris & crew

Stephenson's rear gunner

B-24, Liberator, 223 Squadron, Oulton

F/Lt Stephenson & crew

Hanger staff with effigy of Lord Haw-Haw. VE

When we drove into the camp through the dispersal area all we could spot were black painted Fortresses sporting an RAF roundel – not a Liberator to be seen anywhere. What was it all about? A few days and lectures later we were much wiser. We would be trained up to Bomber Command standard and take the role of airborne jammers to counter the new German V2 rocket. There was much to do and learn in this newly re-formed 223 Squadron. It had been disbanded in Italy and re-formed into a South African squadron after taking part in the desert campaign flying Baltimores.

Now we were in 100 Group Bomber Command. High altitude flying was something we had never done before. All the crews went to the decompression chamber. I went in with a heavy cold, something not recommended, and finished up a day later in Station Sick Quarters with sinus trouble. After 14 hours and 20 minutes of flying training we were considered ready for our first op on 3rd October 1944. It did not start off at all well. Myself and two others were a little late for the briefing and our pilot, P/O Briscoe, played up hell. He threatened us with a court martial if it happened again. We were never late again after that telling off. The briefing intelligence officer was very casual about it all as by this time several of the crews had already flown this anti-V2 patrol.

Taken Aback by Flak

'You will have two Spitfires for escort', said the officer, 'Patrol between Scheldt Estuary and The Hague. No flak or opposition has so far been experienced'.

'A milk run', someone said.

Despite all those assurances my heart was in my mouth. I was scared stiff. There were 12 of us in the crew and I wondered if the others felt as I did. When we had flown the Liberator in Nassau we'd had only seven in the crew and there was lots of room, flying in just shirts and shorts. Now, with our thick underclothing, battle dress, electric suits, Mae Wests and parachute harnesses, space was at a premium. Much of the flight deck had been allocated to the front special operator and all his jamming gear.

On the way out our Spit escort overtook us and that was the last we saw of them. Soon we were in view of the Dutch coastline and had hardly completed the first leg of the patrol when all hell erupted. The flak burst very close to our wing tip, too close for comfort. Our mid-upper gunner constantly monitored the sky above both for fighters and V2 launchings. Every time he turned the turret he 'clonked' the wireless operator in his back. This eventually led to a break in the oxygen bottle feed line, and for the rest of the op I had to constantly supply the gunner with a recharged portable emergency bottle.

I suppose we had done about 15 minutes of our four-hour stint on patrol when we saw our first V2 vapour trail; it was going upwards at a hell of a lick. Our wireless operator sent a signal that we had seen a launch. The special operator threw all his jamming gear switches, but it made no difference – on and upwards the V2 went.

Next stop was debriefing, by the same intelligence officer. He asked a lot of questions. I am quite sure he did not believe our story about the flak and had convinced himself that we were 'line shooters'.

'Well, how did you get on, Whizz?' This was a nickname given me by my crew.

'Well, alright,' I said, 'but we had a lot of flak burst very close to us soon after we commenced the patrol'.

My fellow billet pals had already completed one or more of these V2 patrols without incident, and were convinced I was not telling the truth.

'You bloody liar! Your very first op and you're shooting a line!'.

They believed my story a few days later, after some of them had flown another V2 patrol. They came roaring back to base with the blood wagon waiting for them at the end of the runway where they pulled out the navigator with a lump of flak in his bum. He was carted off to Norfolk & Norwich General Hospital to have it removed; it had broken into nine pieces. Vic Green, a fellow flight engineer and a good friend, saw me later.

'Whizz', he said, 'I'm sorry we called you a liar the other day; it was just like you described. We had just started our patrol, when **bang!** – The navigator cried out that he was hit'. Vic said he went into the nose section to try and dress the nav's wound but in his haste to cut off his clothing he forgot to switch off the nav's electric flying suit. There were blue flashes everywhere. Many years later I did some research which revealed that the flak had come from flak ships operating in the area of Nordwal.

More Trouble

Just three days after our first op we were flying our second daylight patrol. This effort did not go off very well either. I would guess that two-thirds of our patrol had been completed when the intercom went dead, and then the radio. I shouted to Briscoe that all three engine generators had cut out and the batteries were flat. He gave me orders to start the APU putt-putt;* I knew that was not feasible. By now everyone was feeling the cold as our electric suits were out of action, plus the mid-upper turret. The skipper decided to abort the patrol. For the uninformed, the prop revolutions were electrically controlled on a Liberator; they had been set at low cruising revs on our patrol and could not now be changed to high revs for the landing and approach. Now we were to see how good our skipper was at his job. With no radio to tell base we had a problem and warn them to prepare for a dodgy landing, in we came and made a perfect touchdown, despite the very low revs. Once again we had seen V2s launched.

I have one small memory of one of our daylight V2 patrols: we had a Paddy for a front nose gunner – a nice chap, but like most Irishmen he liked his beer and liked to bend his elbow most evenings at the sergeants' mess. We had been on V2 patrol for some time when Paddy asked permission to attend to the call of nature.

'Stay where you are, you are not to leave the turret.'

When we returned he was swamped with the remains of last night's beer. No one would sit near him in the crew bus as he smelled strongly of pee. Paddy later became redundant as we dispensed with the front gun position.

On a practice air firing exercise in the Wash area we fired our guns from all positions. It was on this exercise that Lady Luck played a part. On landing we found a .5 bullet hole which had just fractionally missed the port rudder control. Another half inch and it would have been curtains for us all.

On reflection, and looking back over the years, I have often wondered if that is what happened back on that fateful Kingsley exercise when a Liberator disappeared without a distress call and leaving no trace whatsoever. Just one more victim of what later became known as the 'Bermuda Triangle'.

On 19th October we flew our one and only night V2 patrol; it

* Auxiliary power unit, putt-putt, so named because of the noise it made.

took 6 hours 5 minutes and was uneventful. By this time the back-room boffin boys had come to the conclusion that our airborne jamming of V2 rockets was in vain, and it was abandoned. Postwar, it was learned that research had been carried out on the wrong piece of German rocket hardware.

5 Libs Join the 'Window' Spoofers

I have mentioned that when we arrived at Oulton we could see only black-painted RAF Fortresses. These were part of 214 Special Duties Squadron who were deeply involved in a role of 'window spoofing' and 'target support' on behalf of Bomber Command. Now, with V2 jamming discontinued, 223 Squadron Liberators joined forces with the Forts of 214 Squadron and were to assume the same role. The rear half of our Liberator bomb bays had been boarded in so that we could carry a huge load of 'window'. This was made of metallic foil strips which were used to fool and mislead the German radar plotters. Also on board was a load of radio jamming equipment which was used to interfere with the control of German searchlights, flak and night fighters.

On 22nd October we were briefed for our very first 'window spoof' to Essen. Despite the fact that we had five experienced V2 patrols behind us I was as scared as hell as we roared up the runway with a heavy load of fuel and silver foil. It was late in the afternoon and still light; we were climbing through cloud when suddenly we began to ice up; chunks of ice were coming off the props and rattling along the fuselage. The skipper put the plane into a steep dive, heading for warmer air conditions which got rid of our icing problem. He called on the intercom, 'It's OK chaps, look for a hole in the clouds and we will have another go'. This gap was duly found and up we went again to climb to our operational height.

Soon our navigator called to the skipper, 'We shall be 19 minutes late arriving at our allotted point, due to the icing up, and will achieve nothing'. On this advice the skipper decided to abort the mission and asked for a course to base. I felt a tingle of disappointment at this turn of events but I also thought our pilot had made a wise decision. Essen was perhaps one of the best defended towns of the Ruhr; I didn't fancy the flak-producers having us all to themselves. By now it was pitch dark and raining heavily. With still a lot of unburnt fuel and unused 'window' on board the Liberator did not handle at all well. Down we came. We seemed to touch down a bit late on the runway. I could see that we were fast running out of this vital commodity and thought a prang was in the offing when the skipper slewed the plane on to the grass by using heavy braking on the starboard wheel. We came to a stop safe and sound. We were bogged down and had to be towed out, but there was no damage. I began to realise more than ever that I was in the hands of a very able pilot.

AOC's Commendation

On 9th November both squadrons were given a stand down but before the day was out there would be many surprises. There had been dreadful weather with sleet, thunder and lightning. Late in the evening I was in the billet on my own mending a puncture in my service issue bike. All the other inmates of our Nissen hut had left camp after receiving the stand down and had either gone to Norwich or Aylsham or the village pub. Two officers came into the hut.

'Who are you and who is your pilot?' they asked.

'I am Sergeant Anthony and Flight Lieutenant Briscoe is my pilot.'

'Get up to the briefing room immediately, you are on ops!'

That news was a bit of a shock! Somehow all my crew had been rounded up, together with four other 223 Squadron crews. Several crews from 214 Squadron Forts had also been detailed for this window spoof to Saarbrücken. We might have made a mess of our first attempted visit to Germany when we had to abort, but this time there was even more determination to complete the operation. And it was successfully accomplished despite snow and terrible weather conditions. Of the five 223 Squadron crews detailed we were the only crew to get through and complete the mission. For this the Station was congratulated on its effort by a personal message from the AOC. It read:

> 'Operations 9/10th November 1944. The long period of standby of the 9th November was an operational necessity. A good effort resulted and the following congratulatory message has been received. The crews that took part may attach this extract to their Flying Log Books. The extract may not be detached, and inclusion in Log Books must be made for security reasons.'

The message was signed by T. C. Dickens, Group Captain Commanding RAF Oulton: A 1103, 10th November. The AOC's commendation read:

> 'The results of last night's spoof raid were most gratifying. Our aim was amply achieved in that the enemy was induced to react in a very big way indeed, first in the threatened area until he eventually became aware he was being spoofed after the windowers had returned, and then in the Ruhr area where he believed a real raid was to follow the spoof. I know how difficult were the conditions last night, how these were aggravated by last minute changes in the programme. The latter, however, were made to take advantage of the best possible weather conditions on a bad night, as revealed by the stop press weather reports. I congratulate all crews who took part in this difficult operation. Their determination enabled the Group to score a very distinct success. Well done.'

This AOC's commendation came just at the right time and did much to build up our confidence. At this point crews were allocated to a particular aircraft; we got T Tommy TS 526. It was not always practical, however, to fly on ops with it for various reasons such as sickness and crew leave.

Six days after our successful 4½ hour trip to Saarbrücken we flew again on a window spoof to Coblenz. Because of extremely unfavourable weather conditions at base we had to divert to Carnaby, near Bridlington. This was a huge base specifically designed for what we called a 'prangdrome'. It had extra wide and long runways and was ideal for those kites that had a problem with their landing aids, such as no flaps, no brakes or, for some poor souls, no hope. A plane could be bellied in and the lights switched to another runway whilst it was dragged off to make room for the next one in trouble.

Next morning we went and checked out TS 522; it appeared to be OK. There was mud and water everywhere. I had never seen so

1456824 F/Sgt ANTHONY D.A.G. (F/E)

A.O.C's COMMENDATION
ON COMPLETION OF OPERATIONAL TOUR.

"For Meritorious service and good airmanship, in that a full operational tour has been completed without having been involved in any accident or ever having an unnecessary cancellation or abandonment of an operational sortie".

Recommended ....
Wing Commander, Commanding,
No. 223 (BS) Squadron.

Recommended ....
Group Captain, Commanding,
R.A.F. Station, Oulton.

Approved

for Air Vice Marshal,
Air Officer Commanding
H. Q. 100 Group.

many smashed aircraft. Lancs and Halibags littered the place with the odd Mossie here and there. What seemed like hundreds of men were spread all over the airfield busily engaged in repairing and dismantling planes. The place was brimful of aircraft because contingents of the US 8th Air Force had diverted to Carnaby the previous day due to bad weather at their East Anglian bases. We were stranded at Carnaby for nearly four days. Carnaby was bulging at the seams. We had no money; we were not allowed to carry it on ops – no soap, not even a toothbrush. There were big long queues for food and toilets. Sanitation was not very good; sleeping accommodation was also very primitive. I slept in my uniform the whole time; I did not fancy the well-used blankets.

Eventually we got away and returned to Oulton to learn that a US 8th Air Force Liberator, which had been on a 'Carpetbagger' mission and diverted to Oulton, had crashed and burnt out 200 yards off the runway. It seemed that the crew took off after diversion to return to their base when the Liberator began to leave a trail of fuel behind. This was a fairly common occurrence with the Lib and could be very dangerous. It usually happened when a badly replaced petrol tank cap created a vacuum which sucked off fuel which could easily be ignited by the turbo exhaust. Rumour had it that one bloke jumped at very low level as the pilot brought the kite in for an emergency landing. The rest of the crew were killed.

We flew nearly every day, whether on air test, air firing, engine run in, fighter affiliation, etc. It was on such a day that the skipper flew us to Lakenheath Hall and did a few gentle circuits around his ancestral home. Roxy Hart, our co-pilot, was a fearless young daredevil; he pleaded with John Briscoe to buzz the place up, but he was not having any of that.

'What you don't know, and I do, is that there is an Observer Corps post very close to my parents' home. They are bound to report me and it would lead to a court martial for sure'.

Flying over East Anglia during the war was quite an experience. One had to keep a sharp look out all the time as there were aircraft everywhere one looked. Looking down, there seemed to be an airfield passing by every few minutes.

Diversions Mean No Rum

On 20th November we took TS 524 on an engine run in and on the same night took it on a window spoof to Karlsruhe. On 23rd November, in TS 533, we carried out another spoof raid, this time to Cologne. The trip took 4$^1/_2$ hours. We used TS 533 again on the 27th when we spoofed Mannheim – a round trip of 6 hours 25 minutes. By now we had completed ten operations. My confidence began to grow about the possibility of completing a tour of ops. Oh well, just another 26 to do. Up until now, with two months of operational flying completed, not one aircraft of 223 Squadron had been lost.

We flew another spoof raid to the Ruhr area on 4th December. This was uneventful except that we were diverted to Little Snoring on our return. This was the home of 23 Squadron, Mosquitoes, to whom I had temporarily been attached at Ford in 11 Group Fighter Command way back in 1942. They had left Ford and had completed a tour of duty in Malta but were now in 100 Group supporting Bomber Command on Special Duties. They also liaised with 223 Squadron and carried out the 'fighter affiliation' exercises to give our crews experience of fighter attacks. We flew on a window spoof to Rheine on 5th December 1944 and another on the 9th to Augsburg which took 6$^1/_2$ hours. On our way back from a window spoof to the Ruhr area on 21 December we received a coded wireless message instructing us to land at Milltown.

'Where the hell is Milltown,' someone said. (We were all longing for that cup of coffee laced with rum.)

'It's in Scotland,' replied the navigator.

That was well over one hour's flying time from Oulton. This trip took 7 hours and 35 minutes, our longest operational flight time so far. Being diverted in Bomber Command was not funny; no one was very interested in our welfare. It was difficult to get a post-flight meal and the rum was conspicuous by its absence. After a lot of questions and guidance we did manage to get our breakfast, and it was here that I met a flight engineer friend who had been with me at training school.

'What are you doing flying those black-painted Libs?' he enquired.

We all had instructions to keep our mouths shut; I passed it off and told him straight out that I was forbidden to talk about our role. Even to this day a lot of ex-airmen are completely unaware of the part played by the cloak and dagger boys of 100 Group, Bomber Command.

I asked my friend, 'How do you like flying with the Lib in Coastal Command?' (Milltown was an anti-submarine base.)

'Not so bad,' he said, 'we homed on to a submarine a few nights ago; all hell was let loose; when the Leigh Light came on it turned out to be a destroyer letting go with all its flak.'

The children at Milltown seemed to have access to the camp area; they must have thought we were Yanks as they constantly asked, 'Got any gum, chum?'

At last we got away from Milltown and set course for base, only to receive another radio message to divert to Middleton St George, near Darlington. There was a short bout of swearing; the weather had closed in at Oulton. We thought we had landed at Middleton but on enquiry found we were at Thornaby, south of Middlesbrough. We received a very cool reception at this Coastal Command station. We had to stay overnight and had a hell of a job finding somewhere to sleep.

Flying Officer Steele had been hit by flak on 5th December; on landing his nose wheel collapsed and he skidded for several hundred yards, but no one was hurt. On 7th December two new aircraft arrived. New!? They had already flown 380 hours; that was considered a lot of hours in those days. On 17 December it was decided to discontinue using a front gunner, which left us with 11 in the crew. All the spare gunners were posted, the guns were removed and the turrets doped over. This led to better handling of the Lib from the pilot's point of view.

Our recent diversions to Milltown had left the squadron with just five aircraft and five crews. Our crew's trip to the North Ruhr had been in support of a main force raid on Bonn and Cologne. On 23rd December 223 Squadron managed to get three aircraft off to 'spoof' Frankfurt despite a big muck-up caused by diversions. The success of this window spoof was the greatest yet experienced as the Luftwaffe put up over 400 night-fighters to meet the attack. Two of them were destroyed and two damaged. I might add at this juncture that getting the Luftwaffe airborne chasing non-existent bomber streams in filthy weather, using up precious fuel, and then exposing their airfields to our special night-intruder patrols, on return, must have caused the Germans severe frustration.

25th December 1944. Another Christmas Day had arrived. The

weather was bright with fog in the evening – no ops. I volunteered to help serve Christmas dinner with other NCOs and officers in the airmen's mess. There was a strong feeling of goodwill between ourselves and ground servicing personnel. I had got to know a lot of them very well; this was made easier by my having formerly been ground staff. One of the fitters I knew well, John Welsford. We had been in the same 'rooky squad' at Yarmouth and trained in the same class at Halton. The airmen's mess had been beautifully decorated for this Christmas occasion. Masses of window had been used as the basic decorating material. I remember, I remarked to someone that it would have served a better and more useful purpose if it had been dropped over Germany.

'That's where you are wrong, old chap, all this window is now useless and redundant'.

'Why?' I asked.

'Because Jerry has changed the wavelengths on his radar'.

Subject closed.

They had scratched together some bods to play in a band whilst we served their Christmas dinner. My reward, when I got to the sergeants' mess, was a nearly-cold dinner.

Boxing Day 1944 was an early stand down with frost and snow everywhere. I was determined to get home for a few hours. It really was a White Christmas. Looking from the train window on my way from Norwich to Colchester I could see the snow frozen on the tree branches – a real Christmas-card scene. Up until 27th December there was no de-icing equipment available on the flight dispersal. Demonstrations had shown that it took 6–8 men half an hour to defrost one aircraft.

Our crew took off at 03.30 on 28th December when we flew another window spoof, this time to Frankfurt in support of, and as a diversion for, Bomber Command's main attack on Opladen. Flight Lieutenant J. Brigham's crew sighted a JU–88 on a trip to Frankfurt. The 88 closed to 1000 yards and then broke away. Returning over the battleline, Brigham and crew were hit with our own flak. The following night we flew to Münster and the night after that to Wesel to support a heavy attack by the mainstream on Cologne. This made three nights out of three that our crew were operational.

On the trip to Wesel we had taken off at 18.25 and landed at 23.20. We dropped window from 20.28 until 21.08. Our 'Jostle' was jamming from 20.28 to 20.55 and then became unserviceable (U/S). 'Carpet' jammer was operative from 20.33 to 21.19 on 10 frequencies. Piperack operated from 20.26 to 21.25. The Monica set became unserviceable (U/S), Monica was a device operated by the radio operator to monitor our aircraft's defence. It comprised a small round screen which was supposed to show a blip if an enemy night fighter was in close proximity, but more often than not it was a useless tool as it was prone to jamming and 'grassed' up. Our wireless operator (Wop) switched it off without the skipper's permission and as a result of this disobedience we were looking for a new Wop the next day, whilst our Wop was sent to 'Sheffield' for his sins. Sheffield was a disciplinary depot where aircrew were sent after falling foul of orders and squadron commanders. Wing Commander H. H. Burnell was our squadron commander, a Canadian who had gained a lot of experience in Bomber Command. He often flew ops in the co-pilot's position and flew with our crew on several occasions, but he did not handle a Liberator as well as John Briscoe. In the early postwar period Wing Commander Burnell was killed when his Lancaster crashed into a mountain in Italy. Sadly, our co-pilot, Roxy Hart, suffered the same fate in 1959 when his Anson hit a mountain in North Wales.

One of the nice perks that RAF operational aircrew enjoyed was seven days' leave every six weeks plus a free rail warrant. I thought this privilege was viewed with some envy by non-aircrew. We always seemed to get away OK when our leave was due. The worst part was that we always found ourselves on the Battle Order the very first morning back from leave. It used to have a slight shock effect. Another perk that aircrew enjoyed was that we were paid four shillings a day extra whilst on leave. This money did not come from RAF sources but was derived from a Trust Fund set up by Lord Nuffield. The original concept was to provide Commonwealth aircrew with a special payment. I never did learn how we English benefited from this favour.

I had two pay rises in December and was automatically promoted to flight sergeant which meant that another one shilling a day was added to my basic 12 shillings a day. I also received my first good conduct badge; this made a grand total of 13 shillings a day. When I had first joined up my pay was two shillings and sixpence a day, of which I voluntarily allotted a shilling a day to my mother; it was a bit of a struggle to manage on the remaining one shilling and sixpence.

Battle Order for Liberator TT 336 'R', 29/30 December 1944. Window spoof to Munster. Take off: 15.55. Crew as follows:

F/L J. Briscoe	Captain
F/O E. Hart	2nd Pilot
F/O A. Love	Navigator
F/O E. Chandler	S op
F/O C. Payne	S op
F/S D. A. G. Anthony	Flt Engr
F/S W. Bowsher	AG
Sgt K. Stone	AG
Sgt D. Wilshaw	AG
Sgt Chitticks	Wop
Sgt Nolan	AG

This aircraft dropped window from 18.43 to 19.40 and VHF Jostle from 18.12 to 19.55 hours. Piperack same as for Jostle. Carpet jamming on five frequencies: 18.35 to 19.32. Monica as briefed. Mission completed, landed at 21.20 hours.

6 1945 Brings New 'Targets'

So far so good. We had completed our seventeenth operational sortie at the end of the year. We knew it would not be long before we were assigned to a 'Target'. This was a completely different task to that of a window spoof. It involved getting over the target to be attacked just as the pathfinders began to mark it with their pyrotechnics. Our task was to orbit above the Lancs and Halibags of the main force and jam the German radar controlling his flak and searchlights. Also we jammed the ground-to-air radio control of the German night-fighters.

Nearly a Fire

It was around this period of my tour that we had another slice of luck. We were on our way back from a trip to Germany; suddenly there was a shout over the intercom from the two gunners at the waist positions:

'We can smell smoke, something is burning!'

In an instant I knew where to look. I grabbed my torch and made for the bomb bay. There it was – smoke was belching from a thick cable right next to the main fuel lines. I knew that this smoking cable led from the front special operator's box of tricks. I told him to switch off his gear and I stood by with a fire extinguisher, but it was not needed. The cable ceased smoking. Some of my readers may wonder why I was so prompt in finding the source of the smoke. Well, the explanation was quite simple, really. The Liberator was a very draughty plane and any smell was quickly passed from front to rear. I could always smell the fag that Alan Love, our navigator, lit on reaching landfall and the last given course for base. Years later I pulled his leg about this crafty move.

'Yes, Whizz, that's quite true, I also used to take a swig from my hip flask.'

If someone farted on the flight deck those to the rear got a quick passing sniff. When the waist gunners had raised the alarm I knew the trouble had to be between them and myself, hence the instantaneous discovery of the smoke.

On New Year's Day we were on the Battle Order. This was to be our very first 'Target' in T Tommy to Dortmund. A bit traumatic. We orbited above the pathfinders and main force and saw how well the night-fighter flares illuminated the main force just below them. We did not see any bombers knocked down. It all went fairly smoothly and after our allotted time over the target Briscoe opened up the engines and went into a shallow fast dive towards the battleline. After getting to this area we settled into cruising conditions so that we could conserve fuel as a diversion was always on the cards. It had taken $4^1/_2$ hours to complete our first Target sortie.

The following night we flew on ops again, this time to Frankfurt. This window spoof effort was made to aid the main force attacks on Nuremberg and Ludwigshafen with two 'feints' to Hanau to divert the night-fighters and prevent fighter reinforcements. On 4th January we carried out a 70-minute 'splash firing' exercise over the Wash. On the 8th January we had to fly to RAF Manston to pick up our deputy flight commander, Flight Lieutenant H. Stephenson, and his crew. They had landed safely after an op with very little brake pressure. Stephenson was a really nice chap, who already wore a DFC for his exploits serving in Coastal Command in anti-shipping and submarine patrols. Whilst in this role, flying Blenheims, he was an only survivor from the squadron. I got to know him a lot more in the postwar period as he came to live and farm in Colchester, just a couple of miles from my home. He flew over 10,000 hours with BOAC. One day I asked why he had not received an award for his tour with 223 Squadron like all the other captains who had been given DFCs. It seemed that one night he had an argument with 'the powers that be' after he had aborted a malfunctioning op, and they had suggested that he should have pressed on. Stephenson disagreed, claiming he would have put his crew and aircraft in a high-risk situation. As a result there was no bar to his DFC. Stephenson's flight engineer was an 'ex-brat'* and good friend of mine. He died in the early postwar years when he crashed in a Lancaster operating out of Boscombe Down, leaving, like Roxy Hart, a wife and three children.

January 14th was a turning point for both our crew and the squadron; we were sent to fly our 20th operation. At briefing we learnt that we were to carry out new tactics for 100 Group. First we were to fly to Frankfurt on a window spoof and then withdraw and orbit over friendly territory for about an hour; then we were to go back in again and carry out a further window spoof on Münster. Total flying time for this was 6 hours 40 minutes. It was not the weather this time that made us divert to Woodbridge on our return, but a Fortress at base with a burst tyre blocking the runway. Once again we missed our rum ration. Next morning we met the same scenes that we had seen at Carnaby – a graveyard of wrecked aircraft. Woodbridge was another 'prangdrome'. At this stage of the war there were no problems with shortages of aircraft or crews. A 35-minute flight back to Oulton from Woodbridge was greeted with sad news; we had lost our first Liberator on ops. Flight Lieutenant G. Noseworthy was shot down 12 miles south of Antwerp; it was thought to be by our own flak. I would like to point out that this was quite a feasible suggestion. At this juncture of the war air-launched V1s were being fired. Antwerp was in Allied hands and receiving more than its fair share of them. Crews had already been told to give Antwerp a wide berth as the flak gunners were a bit touchy and trigger happy. Noseworthy's crew had been to Gravenbruch.

I have written that 14th January was a landmark; we had completed 20 missions and were now to lose three of our regular crew. Our mid-upper gunner P/O Roberts had finished his second tour in Bomber Command. He had flown his first tour on Halibags as a bomb aimer and had been awarded an MID† for his participation in the Peenemünde raid. After the war at Abingdon there were two WAAFs working in the hangar stores, both wearing 'oak leaves', and I asked how they had come to be awarded. 'Well, we were in the stores nearly all the war,' they said.

* Boy entrant who had been given two or three years' training as an apprentice to his particular trade.

† MID, Mentioned in Dispatches. This award showed a small oakleaf cluster.

I ask you, how does that compare with a man who has bombed Peenemünde? Flying Officers (F/Os) C. Payne and E. Chandler, both special operators, had flown 20 missions with us and had finished their second tour, their first having been on Wimpeys in the Middle East.

Meanwhile, we had word from our ex-Wop. in Sheffield. Dramatic news: he was going to be discharged from the RAF on medical grounds. He was found to have diabetes mellitus. This news came as a shock, but it explained why he had spent time in Oulton Station sick quarters with a facial rash, and why he kept creeping out to slake his thirst in the sergeants' mess. One night he left with a skinful to return to sick quarters only to be bowled over by a WAAF-driven truck. Poor Frank, I heard from him after the war; he had got a job as a wireless operator in the Foreign Office.

Before he had been tipped out of our crew, Frank had quite an unusual experience. On this particular night, prior to ops, we had just begun our taxiing run from dispersal to the runway. Bomb doors open, yours truly keeping a sharp lookout from the top escape hatch above the flight deck, this was standard procedure. The putt-putt was running, together with the electric hydraulic booster pump. Frank was sitting at his table already 'listening out' on his assigned frequency for the night's operation. As I went through the drill and carried out the pre-take-off checks, I noticed he wore a frown and looked worried. After the op I learned why. As we had begun our move to the runway Frank had heard a dance band playing, then a voice proclaiming he had just been listening to Bob Crosby and his orchestra, followed by an announcement that the British Terror Flyers were about to leave their bases en route for Germany. I was glad he related all this after we landed. All the other Wops had also heard this broadcast – clearly a ploy by the enemy propagandists. They also had another little trick which concerned radio transmissions. We had an old clapped-out wireless set someone had scrounged and installed in our Nissen hut billet and we would tune in to what we thought was AFN (American Forces Network). In reality it was a German re-transmission of AFN. Every few minutes the transmission would stop; then the names of British PoWs would be given, including quite a few airborne personnel who had been 'put in the bag' at Arnhem. Cheeky devils.

*

There was a large wood near our Nissen hut which contained a sizeable population of pheasants. I ventured out late one afternoon to see if I could pot one with an airgun I had. I lay low in the bracken, every now and then popping my head up to look for a target. I looked up for the umpteenth time only to see the gamekeeper with another bod only a short distance away. I took off on all fours, squirming my way through the undergrowth. I felt sure they had spotted me but kept going until I finally crashed through a hedge and ditch outside the sergeants' mess. Nobody had seen me come out of the ditch so I bolted into the mess toilets.

I carefully picked off the bits of bracken and grass from my battledress, shoved the gun down my trouser leg and walked in a distinctly lame manner back towards my hut. I had got away with it. Oh no I had not! Walking towards me was the gamekeeper and his companion; the game was up. I could see a court martial or a charge looming on the horizon. I knew that the landowner was thick with the station commander. 'Brazen it out', I told myself, so I hobbled past the two of them, casually saying, 'Hello'. Further along the path I limped past two RAF bods and overheard one say to the other, 'He walks as if he has done something in his trousers!'. I had literally come very close to doing just that. I kept out of the woods after that episode.

On the 1st February we had a distinguished visitor to Oulton, Air Vice Marshal Sir Hugh Pugh-Lloyd, who showed a keen interest in the operations carried out. The next day we had another visitor, not so important but more than welcome: Sergeant G. Mellors who was the sole NCO survivor of TT 336. He thought he had been shot down by a night fighter, which only shows how confused and contrary reports could be. In the first instance it was thought it was our own flak. I am convinced, even to this day, that many shot down airmen never knew what hit them, particularly 'Schrage Müsik'* victims. We were unaware of this deadly upward-firing cannon, perhaps this was just as well. What you don't know you don't worry about. Four days after the return of Sergeant Mellors his skipper, Flight Lieutenant Noseworthy, showed up. Their nine deceased crew members are buried in the British Military Cemetery at Hanover. The remains of their aircraft is exhibited in a 'Wings of War' museum at Meerhout, Belgium, close to where it had crashed.

* Schrage Müsik (slanted music): upward firing cannon.

Solid as a Rock

From now on we had to find spare bods to make up a full crew. Both the Gunnery and Wop leaders flew part of their second tour with us. We never had any trouble making up crew numbers, Briscoe having made a name for himself for being reliable. To me he was solid as a rock; I would not fly with any other skipper. In fact, if the engineer leader was looking for a spare flight engineer I made sure by keeping out of his way. F/O Barratt was engineer leader, an Irishman with a long experience of Liberators in Coastal Command. I can only remember one flight engineer at Oulton being 'tipped out' of his crew. His crew had been on an op and on return he could not get the nose wheel down. They circled base for a while, during which time they stripped one of the waist guns and used the barrel as a fulcrum to try and lever out the nose wheel from its well. Finally they came through on the radio asking for help with their problem. The flight engineer leader had to give instructions. Simple really, one had to pull back a spring loaded lock and then the wheel could be lifted from its housing and thrown out. Next day the flight engineers all got a strong lecture from Paddy about nose wheel emergencies. The guilty flight engineer found himself without a crew next morning. He was moved back into the training flight.

Our next operational sortie on 16th January 1945 was to Flensburg, near the Danish border. It was a fairly straightforward window spoof but on the way back there was a little confusion due to the skipper getting a wrong course. The flak hosed up to make things worse and it was then realised that it was coming from flak ships in the harbour at Wilhelmshaven. T Tommy TS 526, our own allocated aircraft, had a No. 3 engine change on 22nd January; in the afternoon we flew it for $1^1/_4$ hours on an engine run-in. Roxy Hart was not around so I got the co-pilot's seat. There was a lot to think about and it got to my nerves because I made a foolish mistake. I left No. 3 engine in *automatic rich* as instructed, but instead of moving one, two and four to *automatic lean* I moved them to *automatic cut-off*. The nose immediately dipped. The skipper knew what I had done, knocked my hands away and reset the controls to *rich*. After a post-flight telling off and an apology from yours truly I was still on speaking terms with my skipper. I could easily have lost my place in the crew for this silly slip.

That same night we flew another long 7 hours 20 minutes in T Tommy; first we went to Gelsenkirchen, then we returned to orbit over friendly territory. We then made a second window spoof, this time to Mainz. Although we had carried out two penetrations in one night we were only awarded one op, but we knew our tactics were to wreak havoc on the German defence system.

One night I found I was the only occupant of the hut and was woken by a hell of a lot of banging. I crept outside and soon guessed what was going on. It was a crashed aircraft and I could see flames and smoke through the trees and tracer shooting across the skyline from the exploding .5s. There was nothing I could do, so I crept back to my bed. In the morning we saw the remains of a 214 Squadron Fortress partially burnt out with its tail in the air. It had been brought back to Oulton after an ME–110 attack. Several of the crew were injured and the navigator subsequently died.

We received revised orders with regard to the number of sorties to be flown for a 'tour of ops'. Ex-operational Coastal Command bods were to fly 30 ops for a tour with 223 Squadron; if 36 they were to be credited with two tours. Others doing their first tour were required to fly 36 sorties and 14 more to call it a Bomber Command 'double tour'. Those with one Bomber Command tour to their credit and on their second with 223 Squadron had to finish at 20 sorties.

We flew another 'Target' support sortie when the main force attacked Hohenbudberg on 9th February 1945. It was a late take-off at 03.27 and we orbited the target area from 06.10 until 06.20 at 21,000 ft. There was accurately predicted flak in the target area, about the same height as our aircraft, and quite close. We had gone through a belt of searchlights before reaching the target. It was the first time we had been hit by flak when a lump passed inches away from the rear gunner's head. Very soon afterwards he reported that his turret was out of action, which suggested to me that we had been hit somewhere in our hydraulic system. The rear gunner continued to use the turret by hand manipulation – not very efficient but better than nothing.

'Get down to the bomb bay, Whizz,' Briscoe ordered, 'Check the hydraulic reservoir'.

I was down there like a shot. The reservoir was at its specified height, so we had lost no fluid. I flashed my torch all through the fuselage. Not a leak anywhere. I told the skipper I thought No. 3 engine hydraulic pump was u/s and he agreed with my diagnosis.

By now the starboard main wheel, with no pressure to keep it up, slipped from its housing and hung half-way down. We decided to leave things as they were until we circled base. By now it was light and we were above the Channel, flying over a friendly destroyer just coming out of Dover. This was my first ever view of those 'white cliffs'; nobody was more pleased than I to see them. Over base I switched on the electric booster hydraulic pump which started with a friendly screech. I opened the emergency cock leading to the main system and we were back in business. Our skipper selected wheels down and out they came.

I nipped up to dispersal later in the day; by then a new hydraulic pump had been fitted to No. 3 engine. Just as I thought, the drive shaft had sheared off – a specially waisted shaft which broke off when there was malfunction of the pump. Strange it should choose to go u/s just as we collected our first piece of flak. Only the flight engineer on our 223 Squadron Liberators wore an American-made Mae West. The size of the British version was such that it was nearly impossible to get through the catwalk wearing one. We needed to get through the bomb bay quickly, as often we had to go to the rear bulkhead to transfer fuel from the Tokyo tanks to the main tanks. Flight engineers also had to help drop the window Type MM and N3,* and be the most mobile member of the crew. When I dropped the window the rear special operator joined me in the bomb bay and gave me a hand. We were given instructions by the navigator when to start and cease dropping. He also gave us the rates of bundles per minute, usually about 10, increasing to 50 bundles, and then the rate gradually tailed off. We had lots of anti-flak and anti-fighter window on board as well. I put this generously into the dropping chute hoping some lurking fighter would go after it on a wild-goose chase. I am not sure if this type of 'chaff' did a proper job, but I know it did much for my morale. The bomb bay was a rotten place to be, with –40° F temperatures. The cold draught was deadly; goggles were essential to stop the blinding tears. After a while one wore a 'bib' of ice, caused by moisture from the oxygen mask. On one such night in the bomb

* Basically, Window Type MM was used to cover 65 to 200 MHz 'Freye' and FUG 220 German radars. Type N3 was used against German radar Wurzburg operating in the 350 to 600 MHz range.

bay my intercom control switch froze solid, which in turn partially blacked out the remaining intercoms.

'Get off the bloody intercom whoever you are,' Briscoe called.

Time on an op went very quickly for me; I had a lot to do and think about. I had to keep an updated fuel consumption log, together with all engine temperatures and pressures, plus a call-up of each crew member about every 20 minutes for an oxygen supply check. Without oxygen a man could die in a very short time. Electric suits were a godsend. I have often thought how grim it must have been for crews earlier in the war before these suits were introduced. They were not, however, 100% efficient; quite often a slipper or glove would fail.

There were no main force operations on 10th February but 100 Group aircraft operated a window spoof in conjunction with a 'Mandrel' screen together with our roving long-range fighters. All this effort was to force enemy night-fighters to fly large numbers of abortive sorties and to frustrate the Luftwaffe and use up their, by now, precious fuel. Mandrel was used to blind the German plotting service. On occasions, faked breakdowns were staged to 'show' an approaching bombing force consisting of a few aircraft dropping window. The planners were always scheming new tactics to fool the enemy plotters.

P/O Wallace returned to the squadron on 11th February, having recovered from the flak injury he'd received way back in October. This was the first time I had heard of a new drug, Penicillin, which had been used to treat him in hospital.

Our crew flew a window spoof to Bonn, a lucky 13th of February, which also proved to be a most successful night for Bomber Command. The squadron covered the heavy bomber attacks on Dresden and particularly the dangerous Ruhr. New records were set up: the squadron despatched 11 aircraft and not one aborted. Two aircraft flew the southern spoof and the other nine the northern spoof.* It proved to be an unlucky 13th February for the people of Dresden, when, in two separate waves, a total of 805 aircraft delivered 2,659 tons of bombs. The following day it was attacked by the USAF and more than 25,000 people met their death. It had been our 24th mission. Strange now, looking back, how that number 24 kept recur-

* Generally speaking, a spoof against the Hamburg/Bremen area indicated a northern attack by the main force. Likewise an attack against Pforzheim and Friedberg would indicate a southern attack by the main force.

ring. Perhaps also a lucky number; the last two figures of my service number were 24, I had got engaged on the 24th, and I was, luckily, still flying and surviving on B.24s. My son was born on the 24th of the month in 1951.

Boomerangs, or abortions, were quite a common occurrence during the war. Lots of faults and problems did not develop until after one became airborne, then it could be anything. There was so much that could go wrong. There was even a case, one night, of a crew having its guns fire by themselves. More often than not there were extreme weather conditions, and human error played a part. Many times we were briefed to fly an op only to have it 'scrubbed' at the last minute. Sometimes we were as close to an op as being on the point of taxiing out from dispersal.

We did another window spoof, this time to Mainz, on 14th February. Airborne at 21.55, we returned at 02.45. This spoof was in support of the main force attack on Rositz and Chemnitz. Berlin was also attacked that night, and 50 aircraft laid mines off Sassnitz. Flight Lieutenant Stephenson had to return early; he had a runaway propeller on No. 2 engine. Our next op was a Target Support sortie with the main force, covering them for an attack on Gravenhorst, close to the Dortmund-Ems canal. Soon after we began to orbit the target the Master Bomber abandoned the mission at 01.04 as the cloud blanketed the target right down to 500 feet. The ground defences did not react to our presence. One enemy aircraft was seen. We returned to Oulton at 02.45 to learn the sad news that F/O Thompson had been shot down in the vicinity of Dortmund, flying J Johnnie. I had known his flight engineer quite well. Flight Sergeant Wynn was a young, happy-go-lucky chap, similar to most aircrew I knew. The following night we were sent on a spoof to Düsseldorf.

On 23 February we were assigned the protection of the Main Force of 374 bombers who were to attack Pforzheim. We used our Jostle, Piperack and Carpet, flying at 12,000 feet. We saw three aircraft go down in less than five minutes. There was just a flash of tracer, next a flaming streak, followed by a fireball bursting on the ground. The waist gunners called out that they had seen a couple of parachutes go down. I was more than pleased when the navigator gave us a course to steer for base. En route we were diverted once again to Carnaby. This had been our longest operational flight so far, 8 hours 10 minutes. As usual we had to be debriefed before we could get our heads down. The intelligence officer asked us about the flight etc., and we told him we had seen three shot down. He did not seem very convinced and implied that we had probably seen 'scarecrows'. These, he said, were shot into the sky by the enemy defences to simulate a shot-down aircraft. Bill Bowsher, our Canadian waist gunner, had the final words on this matter. 'Oh, is that so, and have they made them so that men jump out of them at the end of a parachute?' Bill never said much; he was a dry old stick with a great sense of humour.

The Master Bomber at Pforzheim was Captain Edwin Swales of the South African Air Force. His Lancaster was hit; he won the Victoria Cross for his bravery in staying at the controls whilst his crew escaped. An American lady of my acquaintance has been researching this attack on Pforzheim which had been the most heavy and concentrated of the war. Two Halifaxes of 192 Squadron, Foulsham, who were engaged in 100 Group's electronic jamming war failed to return from Pforzheim.

Bomber Command did not carry out any major attacks on 24th February, so 223 Squadron backed up a force of 90 aircraft derived from OTUs as far as 05 degrees east, and then flew on, leaving the OTU stream to return to their bases. A 214 Squadron Fortress was lost on this operation. So also were 4 Halifaxes from 462 Squadron RAAF operating with 100 Group out of Foulsham.

A Tribute to the Aussies

Many years have passed and many of us have forgotten the efforts and sacrifices that were made by Australians in the Second World War. Maybe a glowing tribute, discovered and requoted by Bob Givens, sums it up:

> 'What manner of men were these who chose to venture half-way across the world to go to the aid of the mother country, defend the freedom of their own and fight for a peace they did not live to see. They were a perfectly ordinary generation of young Australian men, little older than schoolboys, who were to perform quite extraordinary feats of heroic daring and to display a selfless devotion to what they saw quite clearly as their simple duty'.

There were no major operations on 25th February but 63 Mosquitoes attacked Erfurt; 223 also had a stand down which was continued into the next day. During this period there were air-to-air and air-to-sea gunnery practice, air tests, and night fighter affiliation training. The weather was good on 27th February; despite a full moon there were no major raids. Because the enemy were becoming used to feint raids on the Ruhr by moonlight, the northern window spoof was sent much further north to Sylte. Eight Liberators took part without incident. On 28th February Bomber Command Mosquitoes attacked Berlin, Nuremberg and Munich. There was minelaying in the Kiel Canal. The American 492 Bombardment Group also laid on an attack on Freiburg; 223 Squadron sent seven Liberators on a window spoof in support of this operation. One of the Libs had to abort owing to engine trouble. Wing Commander Burnell led the formation, flying with F/O Young's crew in P Peter. During February the squadron had carried out 33 Target and 88 window spoofs – a total of 121 sorties. Only one Target and five spoof raids were aborted. Oulton had only five stand downs during February and had operated on 18 out of 28 nights, including seven in a row, 18th–24th.

Berlin, Erfurt, Oldenburg, Bremen, Hamburg and Lübeck were all attacked by Mosquitoes on 1st March. Two Liberators which formed part of a window spoof were routed through Kaiserslautern and then fanned out to suggest an impending attack on Mannheim. The weather was fine on 2nd March. No major raids took place but another window spoof was laid on in conjunction with 70 OTU aircraft which they supported part of the way to Bremen. N Nuts had to abort with engine trouble.

The night of 3rd March 1945 was a dramatic one for Bomber Command. Kamen and Ladbergen were the targets for the main force whilst Mosquitoes were to bomb Berlin and Duisburg. There was also a programme of minelaying. All went well until their return when enemy night-intruders attacked the returning bombers. B Baker of 223 Squadron was in the circuit and about to land at Oulton when air tracer was fired at it, but no strikes were seen. Minutes later 'J', a Fortress of 214 Squadron was shot down and crashed close to Station Sick Quarters. P/O Bennett* was killed with most of his crew; the two waist gunners survived. Bennett had recently been commissioned and had only held the rank for two days. There was further ground strafing during which the

* P/O Bennett was the victim of Lieutenant A. Doring, flying a JU 88 of 10/NJG 3.

Intelligence and Signals sections were both struck and some personnel were wounded. The squadron commander was extremely lucky – he just missed being struck by cannon fire as he left the Intelligence block. B Baker quickly flew off and landed at Woodbridge. It had had a miraculous escape; when inspected it was found to be Category AC. The JU–88 had severed the starboard aileron control and put a large hole through the propeller blade of the starboard outer – also a hole in the starboard wing. F/O Ball and his crew had their share of luck on that night when up to 70 enemy night intruders slipped into the returning bomber stream and struck in a wide area of Norfolk to Yorkshire with considerable success. 'S' and 'G' were diverted to RAF Brawdy in Wales; G was flown by Flight Lieutenant Tony Morris, an American in the RCAF. Recalling the intruder episode some 45 years later, his Wop, Jim Bratten, Jim told me that they returned to Oulton from Brawdy the next day. There were several Canadians in his crew; they went to the Intelligence Block for de-briefing. They were asked if they had seen anything of the night intruders.

'No, sir,' quipped one of the Canadians, 'but it looks as if you did,' as he pointed to the cannon shell holes. Those Canucks were full of fun.

Most of F/O Thompson's crew were Canadians, nearly all from the same billet. When they'd been lost, back in February, someone had chalked 'The Morgue' on their hut door. Usually it was the adjutant's task to collect and return personal belongings. A common sight on railway station platforms in those war days would be a box, about tea-chest size, painted black. On its lid would be painted in white an RAF number, name and rank, stating that it contained personal belongings and addressed to next of kin. A sad sight. Seeing these boxes reminded me and my fellow aircrew that any day in the future that was how our own few belongings could finish up. One did not dwell on this subject for very long. Most aircrew managed to fix their thoughts round to the attitude of, 'It will be someone else that gets the chop, but not me'.

Adopting Plan B

Lucky me again; I was on leave when those Intruders had attacked Bomber Command on 3rd March and created so much havoc. Our crew had developed a nice habit of keeping out of trouble. So far we had only had one 'Boomerang'; some crews had aborted several times because of various malfunctions. Our luck could not last forever. One evening towards the latter part of our tour we had been airborne for a very short time, en route for another window spoof. It was still light and everything appeared to be going smoothly. A call came on the intercom from the rear special operator.

'Skipper, I am not well.'

'Can you carry on?'.

'No sir.'

'Navigator, give me a course to steer for base, we are going to abort.'

'Oh, bloody hell,' someone said. It was awful, after getting airborne and setting one's mind to flying the op to have to abandon it.

From out of the blue I had a bright idea. I suggested to Briscoe that we should send the front special operator to the rear as replacement, and that I could operate the front jamming equipment. I was capable of this task as it was only a case of throwing certain switches at a given command by the navigator. I had watched the special operator so many times over the last few months that I knew I could do it. I managed to convince our skipper; he opened up the engines and steered for base at high speed. We arrived in the gathering dusk, kept radio silence and made a good landing despite still being heavily laden with fuel and a huge load of window. I swear we broke all world records for speeding on a perimeter track and were soon close to the control tower. We pulled up and I quickly left the plane with our sick crewman. I saw him safely to the edge of the perimeter track then sprinted back to the Lib.

'OK, let's go', I called.

In no time we had reached the runway, got a 'green' and roared back into the sky. The navigator had worked hard; he gave a new course which cut a few corners. All went well, we carried out our assigned task successfully and on time too. We came back from that op feeling great, but apart from Briscoe no one said, 'Thanks'. That is how it was, though. Everyone was taken a bit for granted. I am sure that there are not many aircrew who took off twice for one op in WW2. However, this episode did have a final sequel some 36 years later when I had a meeting with my pilot and navigator at a Bomber Command reunion at Grosvenor House, Park Lane, London. Both had done well for themselves in civvy street. Sir John had become deputy commandant of Heathrow airport in 1959 and Alan Love had become an air traffic controller at West Drayton.

One can imagine there was lots to recall and chat about.

'What about that night when we took off twice; I never did hear what was wrong with that sick special operator?' I said.

'I did,' replied John. 'He claimed he was suffering with sinus trouble, but really he was LMF (lack of moral fibre). He had been put in the decompression chamber the next day to test his so-called sinus problem. He was tested first at low levels, then without his knowledge they took height conditions to 18,000 feet and, when asked how he felt, he said "OK". Of course, if he was really having genuine sinus trouble he would have been quite ill. His bluff had been called, the game was up, and like most LMF cases he was quickly posted. He had done 30 on his first tour and was on his 17th in his second tour with 223 Squadron. I suppose his nerves had got the better of him.'

'Why did you not tell us at the time?' I asked.

'What you don't know about you don't worry about,' John replied.

Now where have I heard that said before? In conversation even more years later, John let slip that he was partially colour blind. Over 30, he had not come into the RAF until 1942, due to a reserved occupation with the Air Ministry.

'How did you manage to pass your aircrew medical with suspect vision?' I asked.

'Quite easy Whizz,' he laughed. 'My father knew all the right people on the medical selection board. It was just the greens and blues I was a bit weak on'.

The old devil. Actually, he was a charming man; to us 'boys' he appeared as a fatherly figure. His discipline was rigid and his example in leadership magnificent. It was his efforts that knocked us into an efficient crew.

We were able to fly our own T Tommy on 7th March when we had a successful 6 hours 20 minutes flight to support the main force in attacking Hemmingstedt. Quite often we flew in the main bomber 'stream', depending on the night's tactics. I never liked flying in those dangerous conditions. Pitch black, one could feel the prop-wash from the kite in front but could not see it. It was just like driving a car at night with no lights and no brakes and with all the possible consequences. Many of Bomber Command's casualties were the result of mid-air collisions.

During our crew's last trip to Hemmingstedt six other Libs completed a southern window spoof. 492 Bombardment Group USAF supported the effort by bombing Dortmund. Another 214 Fortress was lost carrying out a Jostle jamming sortie over Harburg. There was little fighter activity. The 12th March saw a massive attack by Bomber Command in daylight. By the end of the day 2,390 planes had dropped their loads on the enemy.

On 13th March our crew flew on a window spoof in T Tommy to Frankfurt but on return we had to divert to Swannington, Norfolk. We took T Tommy on another window spoof on 16th March, this time to Hanau. The squadron received a new aircraft from Burtonwood near Warrington which had already flown 370 hours. The weather had deteriorated on 17th March so operations were scrubbed, but we were up again on the 19th for another trip in T Tommy to Kassel. There were two long targets on 20th March when 100 heavies were to attack Heide and 200 heavies went for Böhlen. The squadron backed up this effort with eight aircraft on a window spoof to Kassel. That night, we lost our T Tommy flown by F/O Ayres. F/O Kingdon from 214 Squadron also failed to return. I felt a pang of sadness and annoyance at losing T Tommy. It is surprising how one becomes attached to an aircraft. It did not do to dwell on this subject for very long.

Operations were planned for two targets on 21st March: Langendreer and Hamburg. Eight aircraft operated to make a northern and southern spoof; one of these was hit by flak but returned safely, but 214 Squadron lost another Fortress, this one flown by Flight Lieutenant W. Allies. There were no main targets scheduled for 22nd March but we were one of seven crews briefed to carry out a window spoof to Bremen. The Mossies did their regular run to Berlin, also Bochum and Paderborn. Bomber Command took full advantage of good weather conditions on the 24th. The night's offensives were co-ordinated with the ground attack forces. Our crew went to Wesel and saturated the Rhine frontline with window. Five out of six planes completed. One returned early with an undercarriage problem. There was very little flak and no fighter activity. The weather became foul on 25th March, and a stand down was announced. A new aircraft was awaiting collection from Burtonwood but the weather had the last laugh. It was learnt that the replacement in question was not so ancient as its two predecessors. Next day there was a further stand down; only Monica training and night fighter affiliation was carried out. A small-scale operation was flown on 27th March with just two aircraft on a window spoof to Bremerhaven. This attack included incendiary bombs and flares and proved successful in conditioning a new area for Bomber Command attacks. 462 Squadron RAAF supplied the pyrotechnics and bombs on this occasion, which caused three *gruppen* of night fighters to take off. A complete stand down occurred on 28th March due to heavy cloud conditions. Only training took place on 29th March. Four aircraft flew a window spoof to Bremen and Hamburg on 30th March.

April began with rotten weather, but on the 2nd the station was able to put up ten Liberators and four Fortresses to carry out a large window spoof on Hamburg. This operation took us just 4 hours 55 minutes to complete. It was estimated that three *gruppen* of night-fighters had been raised on a 'fool's errand'. We used VB 852 to fly on the Hamburg 'feint'. I believe this was a J model Liberator. Next day the bad weather put a stop to flying. However, on 4th April the squadron put up 12 Liberators, all of which completed their missions. Two aircraft were allocated 'Target' support missions, one on Leuna, the other to Lütgendorf. One of the window force had to land at Juvincourt, France, owing to a petrol leak at 01.33. A total of 48 aircraft from 100 Group had taken part in this latest Bomber Support operation.

There was little hope of flying on 5th April, due to heavy rain. We had a visit from Squadron Leader Wills of 100 Group HQ. He carried dramatic news: first, that 223 Squadron was to be converted to B17 Fortresses; secondly, that the 50 ops required for a double Bomber Support tour were to be reduced to 45 ops. Immediately three crews were detailed for Fortress training in 1699 Conversion Flight. It was around this time that we had a letter from Warrant Officer Cole, a special operator who had failed to return from ops on the night of 20/21 March 1945. He was the sole survivor of F/O Ayre's crew. In his letter, pinned up in the locker room, he described his last moments within T Tommy. He was flying as rear special operator when the plane was attacked. He found himself pinned down and unable to move owing to the force of G, and resigned himself to crashing with the burning aircraft which eventually came down in a heavily forested area. His German would-be rescuers could not manage to remove him from the wreck as he was straddled by the heavy Jostle jamming set. Eventually, block and tackle lifting gear was used to free him from his plight. He had been taken to hospital with severe back injuries, and after a few days the hospital was overrun by advancing American ground forces.

Some Thoughts for Future Historians

At some time in the future historians will be bound to ask, 'Why were Liberators and Fortresses used in 100 Group?'. Already some of them have put forward their pet theories; most of these, however, do not hold a lot of water. I offer my own ideas on this subject.

Both Libs and Forts had a high surface ceiling which was always higher than the Lancs and Halibags. We needed that extra height when we orbited above the target and the attacking Main Force. Fitting a Jostle set into a Lanc or Halibag would have been an extremely difficult task. The Jostle set was about the size and shape of a dustbin and a half, and quite heavy, 600 lb. Getting it into a Lib was a piece of cake. A deep concreted pit was dug at dispersal with access steps. The Jostle set was lowered into the pit, and then the receiving Lib was manoeuvred over the pit. The Jostle was pushed into the aircraft via the space left by the removal of the vent ball turret which, incidentally, played no part in 223 Squadron Libs' defensive armament. Using the 'pit method', installation and removal of Jostle was carried out fairly quickly.

Further reasons can be found for the use of Libs and Forts in the window role. Our fuel capacity would allow us to support any target and spend time in orbiting and protecting the Main Force. With four gunners and six .5 guns between them, the aircraft – a crucial part of the night's operations – was better defended than the Main Force Lancs and Halibags. Some authors have suggested that Halifaxes would eventually have become the standard four-engine aircraft to be used in 100 Group future operations. This opinion must have been based on rumour only. Only the Fort and Lib were capable of keeping above a target in 1944/45.

Another question often asked: 'Why, in the last weeks of the war, was a decision made to re-equip 223 Squadron with Fortresses?'

Perhaps forward planning played a part here. With two Squadrons of Fortresses at Oulton, better utilisation of both ground and aircrew could have been achieved. Spare parts and stores could have had a streamlining effect. In retrospect, looking back on this point, I believe the planners were thinking of the forthcoming aerial attack on the Japanese mainland. There is no doubt that if the attacks on Hiroshima and Nagasaki with atom bombs had not taken place, then 100 Group would have moved to the Far East, lock, stock and aircraft.

Why was a decision made to standardise on Forts rather than Liberators?

Another good question. I offer my own personal answers for what they are worth. Taking into account my previous suggestions that 100 Group were destined for service in the Japanese theatre, the Fortress would have been favourite. It had an H2S radar device fitted into the chin turret position, which would have made it a more navigable aircraft around lots of the small islands in the Sea of Japan. Not one of 223 Squadron's aircraft carried an H2S set. This may have been a blessing in disguise, both for myself and my fellow aircrew, as some years after the war I learnt that German night-fighters were fitted with 'Naxos', a device which homed on to H2S; it must have been responsible for sending many hundreds of aircrew to an early death. My observations on the use of H2S carry a little more weight when one considers the difference in losses between the two squadrons. 223 Squadron lost just three Liberators over an eight-month operational period. The loss of 214 Squadron Fortresses was much higher over the same period, at a time when they were carrying out an exactly similar role. Over a period of 6 weeks, between 8th February and 22nd March 1945, eight Fortresses were lost. It is also worth noting that not one Liberator of 223 Squadron was lost in a flying accident. I feel that that is indeed a wonderful record, particularly remembering that the winter of 1944/45 was very severe. Flight Lieutenant J. R. Mcpherson came very close to spoiling this squadron achievement. He was taxiing in F on 29th October when the port oleo was completely sheared off. If he had been landing or taking off there would have been dire consequences. Everybody who was aircrew in World War 2 needed those 'little bits of luck'.

There was a certain amount of frustration amongst our squadron aircrew; flying a 6-hour stint with all its hazards, just to drop a load of silver foil strips? It was for some a large pill to swallow. Probably this is why the odd piece of concrete and bottle were dropped over enemy territory. Even fire extinguishers went missing, and it did not take a lot of reasoning to realise where they too finished up. Postwar, I have often thought with a smile of how a German bomb dispersal squad would have coped with them. Looking for a fuse or detonator must have led to much scratching of the head.

We had another of our aircraft land at Juvincourt on 7th April; this was already pre-arranged before take-off to fit into bad weather conditions. Its pilot, Flight Lieutenant Morris, returned to Oulton next day; his Engineer, Vic Green, told me he was glad to get away from Juvincourt as there were still mines on the airfield left by the enemy.

Targets for 8th April were Hamburg and Lütgendorf. Radio counter-measures were supplied by 223 Squadron, which included three aircraft operating a southern window spoof, and six involved in a northern window spoof. Fog came in at 23.00 hours and closed the base down; consequently all aircraft were diverted. The Flight Commander, Squadron Leader A. J. Carrington, was shot up with flak; P/O Pocock had turned into a hole on a runway when landing under severe weather conditions. These diversions severely interfered with the following night's plans, but with good weather the squadron supported attacks on Kiel and Stade. The AOC sent his congratulations on the excellence of the previous night's operations which were most successful despite bad weather.

Two attacks were planned for 10th April. 250 aircraft were to bomb Plaven at 23.10 hours and at 23.00 hours Leipzig would receive bombs from 70 aircraft. Both targets were covered by 223 Squadron; a window spoof was also supplied. One of our planes was obliged to land in France due to a fuel shortage. Flight Lieutenant Stephenson carried out a Target support sortie in a Main Force raid on Kiel on Friday the 13th. Three other Libs flew a window spoof. Potsdam was the target on 14th April, supported once again by 223 Squadron. F/O Sellars landed at Manston at 02.05; 2 minutes later his port wing was on fire. No one was hurt. Lechfeld airfield received the attentions of a window spoof on 15th April, but it was uneventful. On 16th April our crew were detailed to fly our second longest sortie: an 8 hours 5 minutes trip to jam over that night's target, Pilsen, in Czechoslovakia. Pilsen was where the first golden-coloured lager was brewed. It was also the home of Skoda – a major arms producer for the German war machine. For this raid we were airborne at 23.35 and landed at 07.40; VHF Jostle was on from 03.26

to 04.58 hours, Piperack from 03.27 to 05.10 hours, and Carpet (using 17 frequencies) from 03.34 to 05.11 hours. Time over target: 04.00 hours at 18,000 ft. The attack on Pilsen appeared to be developing well. The target indicators and bombing were well concentrated. We could also see distant flashes which were coming from the direction of Prague, used as a false diversionary target. We were unaware of it at the time, but we were flying our last operational sortie, our 36th.* Next day's surprising news was that a 'tour' had been reduced from 33 to 30 sorties. Whilst we were over Pilsen Flight Lieutenant J. R. Croft had carried out a 'Target support' on Schwandorf.

I was the only member of our crew who never missed a trip with our skipper on his 36 ops.

'Well, Whiz', the skipper said, 'you have got a choice: 6 months' screening, or fly a second tour with me when I convert to Fortresses. Roxy is getting a crew of his own, so you can have the co-pilot's seat.'

I had stood, kneeled, squatted and crouched on the last 36 missions; the luxury of a seat was irresistible.

'Yes,' I said, 'I'm your man, count me in.'

At the same time a little voice from within me was reminding me that it was better to fly with the devil you know than the one you don't. This helped me to make my decision quickly. At this particular time I did not realise how close we were to the end of the war.

223 Squadron history was made on 19th April when Flight Lieutenant G. Bremness flew a 'first' successful window spoof with a B17 Fortress. Three days later Flight Lieutenant Stephenson flew a window spoof into Denmark and in so doing became one of two pilots to complete a double Bomber Command tour.

Before starting our Fortress training our crew had 'end of tour' leave. Operations on 23rd April were a spoof to Luběck via Hamburg with two Libs and a Fortress. Next day four aircraft carried out a window spoof to Munich. There was a stand down on 25th. At this time the first crews for Fortresses arrived from 1699 Conversion Flight. The month ended with a bout of bad weather which included rain, sleet and snow.

With a weather improvement on 2nd May a maximum effort was

* Recent research I have conducted has revealed the coincidence that I flew my first and last operations in the same aircraft, TS 533 'M'.

ordered; nine crews were detailed to fly a window spoof to Kiel, all successfully completed. Ironically, two crews of 199 Squadron at North Creake, in North Norfolk, did not return from this last 100 Group operation of the war. What rotten luck. On 3rd and 4th May there were no operations. On 5th May there was a request for three crews for posting to Transport Command, each crew to consist of pilot, navigator, air bomber and Wop. Flight Lieutenants A. Allnut, Croft and Briscoe were the chosen pilots. No flight engineers were asked for. All thoughts of flying in a Fortress on a second tour had, overnight, 'Gone for a Burton'.

7 No Second Tour and an End to War

There was good flying weather on 6th May but no operations. It looked very much as if Nazi Germany would not be able to hold out much longer, and that night bombing, and hence Bomber Support, had ceased to have any tactical value. On 7th May six aircraft flying in clear blue skies went for a 6-hour flight over the Ruhr area to give crews the opportunity to see by day what they had flown over so many times by night. V-E Day was 8th May 1945, the war against Nazi Germany having come to a conclusion. The joyful news came as something of an anticlimax and, strangely, seemed unbelievable. There were no wild scenes of rejoicing. I found it difficult to realise at first what we had survived over the years. For a number of reasons all aircraft were grounded until further notice. A bonfire party was held, during which effigies of Lord Haw-Haw and Mars, the god of war, were well and truly reduced to ashes.

There was no air activity on 9th or 10th May but there was a lot of ground work going on which produced six aircraft which flew many ground staff over the Ruhr and the Rhine. The route taken was Gravesend, Dungeness, Calais, Aachen, Duren, Cologne, Essen, Dortmund, Dunkirk, Dover, Gravesend, Oulton. All who flew were very impressed by the evident force and destructive power of our air offensive. This trip was repeated next day and was followed by a stand down on 13th May. A Thanksgiving Service was held in a hangar on the airfield on 14th May and over the next five days more trips to the Ruhr were flown.

By the end of May most of the Liberators had been flown to a maintenance unit at Lichfield. On 4th July the last Liberator was ferried to Gatwick by Wing Commander Burnell. 223 Squadron was disbanded on 29th July 1945.

*

How had my three schoolfriends fared – the ones who had undergone training with me at St Athan?

Colin, who flew in Halifaxes, was shot down and buried in Holland. Dick had a mid-air collision with another bomber which he survived. Later his Lancaster was hit by an incendiary bomb which led to his having to jump for it; he finished the rest of the war as a prisoner. Nigel, on Halifaxes, got through his tour safely. Postwar, he became a schoolteacher and, indeed, my son was one of his pupils. Nigel died whilst still a young man.

A magnificent Victory Parade was held through the streets of London on 8th June 1946. After the naval and army contingents came the Allied air forces, led by the Central Band of the Royal Air Force. Bomber Command was represented by Groups 1 to 8, 26 Signals, Numbers 91, 92 93 and 100 Groups, the latter representing the late 223 Squadron. The Parade was accompanied by an associated Fly Past, with representative aircraft from all the Allied air forces.

On May 8th, 1994 it is hoped to unveil a memorial to the men who served at RAF Oulton during the war. It is also intended to place some form of museum in East Anglia which will strive to demonstrate what an important contribution 100 Group made. These projects will need financing. Contributions can be made to the following:

100 Group RAF Oulton Memorial Association.
The Secretary, Mrs. E. J. Boorman,
'St. Clair', 47b Albany Road,
St Leonards,
E. Sussex TN38 0LJ.

Reference

Hillary, Richard, *The Last Enemy.* London, Macmillan, 1942 Pan Books, 1969.

Suggested Further Reading

Carruthers, P. W., *Of Ploughs, Planes and Palliasses – Memoirs of World War II.* Fontwell, Woodfield Publishing, 1992.

Chisholm, R., *Cover of Darkness.* London, Chatto and Windus, 1953.

James, R., *Avenging in the Shadows.* Northampton, Abington Books, 1989.

Jones, R. V., *Most Secret War.* London, Hamish Hamilton, 1978.

Peden, M., *One Thousand Shall Fall.* Ontario, Student Publishing, 1979.

Price, A., *Battle Over the Reich.* Surrey, Ian Allan Ltd, 1973.

Price, A., *Instruments of Darkness: A History of Electronic Warfare.* London, Kimber, 1967.

Streetly, M., *The Aircraft of 100 Group.* London, Robert Hale, 1984.

Streetly, M., *Confound and Destroy.* London, MacDonald and Jane's, 1978.